THE SECRET LANGUAGE OF CATS

THE SECRET LANGUAGE OF CATS

HOW TO UNDERSTAND YOUR CAT FOR A BETTER, HAPPIER RELATIONSHIP

SUSANNE SCHÖTZ

TRANSLATED BY PETER KURAS

Recycling programs for this product may not exist in your area.

ISBN-13: 978-1-335-01389-7

The Secret Language of Cats

Published by arrangement with Benevento Publishing, a brand of
Red Bull Media House GmbH.

Library of Congress Cataloging-in-Publication Data has been applied for

HanoverSqPress.com
BookClubbish.com

Printed in U.S.A.

To Lars, for putting up with my crazy-cat-lady behavior and for loving all of our cats as much as I do.

THE SECRET LANGUAGE OF CATS

PREFACE

WHY IS THE LANGUAGE OF CATS (STILL) A SECRET?

It is a fair question. Since this book presents the sounds that cats use in their vocal communication with other cats as well as with us humans and describes them carefully, and even uses sound and video to clarify, the language of cats is actually not a secret anymore, right? And yet, even after my numerous studies of cat sounds, something still seems to elude me, and remains hidden, like a secret. And is not this last little bit of mystery the reason that we continue to investigate, the reason that we want to comprehend everything a little bit more precisely? For me, at least, the answer is a resounding yes.

Cats express themselves vocally differently than humans. We have to begin by observing their behavior closely so that we can learn their vocal communicative signals and come to understand them as complete beings. We have to crack their "secret code."

We begin by examining the assumption that everyone understands a word in the same way, that everyone defines words identically. But is that really the case? Take the word *yes*. Does *yes* always mean *yes*? Or is it sometimes actually more of a *yeah*? Or even occasionally a *no*? The meaning of a word, what the speaker intends when they say something, always depends on the context, as well as on the speaker's emotional condition or attitude. It's a good thing that if a word produced by a human speaker is unclear, you can always ask for clarification.

What about foreign languages? Well, if I do not know any Hungarian, for example, I can rely on Hungarian dictionaries and on translations. Hungarian has a grammar, and there are books about the Hungarian language which I can rely on for help. I can take language courses at a community college or a university. I can practice with native speakers.

It is different with cat language. Even if I think that I understand a cat sound correctly and can imitate it somewhat accurately, I can never be 100 percent sure that I have interpreted it correctly, whether I am using it in the right context, and how I might interpret it or even try to translate it into a human language. Cats do not have a language that works like a human language.

Even so, we can approach the vocal language of cats and learn to understand it better. The sounds of animals belong to a kind of communication that depends on the situation or context in which the sounds are uttered. You have to study the circumstances of those utterances very closely before you can begin to recognize patterns, let alone a system. In order to study cat sounds more systematically, we can play our cats prerecorded clips of cat sounds and study their reactions very closely. We can analyze the results and interpret the reasons that a specific sound produces a specific reaction.

These are exactly the kinds of studies I conducted with my cats. Although I am pretty sure that the trilling or cooing with which my cat Kompis greets me every morning is a form of friendly hello, I will never be able to enter his vocalization into a dictionary, as cat language does not have words and sentences with a grammar, with structural rules for how to compose words, phrases and sentences—and what these units mean—as is familiar to us from human language.

What does help, if we want to understand the language of cats, is paying attention to the context in which a cat expresses itself. While human languages ascribe identical or similar meanings to different words (a *table* is called *Tisch* in German, *bord* in Swedish and *zhuozi* in Mandarin Chinese), cat sounds always seem to be tightly bound to specific situations. One-to-one translation from human language to cat language and vice versa are therefore impossible. We cannot look something up in Cat. One more reason, then, that the language of cats remains a secret.

What is more, we still know very little about the various categories, subcategories and variants of cat sounds. Most human languages also have variants such as dialects and sociolects that are used within a specific group, sometimes defined geographically and sometimes defined sociologically, e.g. through geographical location, profession or age. These linguistic variants can still be understood, translated and described. Cats, too, may have developed something similar to dialects: in the situations where they have been successful with vocal communication, they will probably continue to communicate with such sounds, and can develop multiple variations (or even learn them from other cats or from their humans) in order to communicate their message more clearly. There are therefore similar sounds that can be distinguished through different vowels or different melodic patterns though they appear in the same context.

Every cat develops, in the course of the life they share with their humans, unique sounds that suit the specific relationship and needs for communication. It is highly likely that the cat has identified sounds that will trigger the expected results from their humans or fellow cats more quickly. Another reason that the language of cats remains a secret is that we can neither precisely interpret, exactly learn, nor perfectly describe these sounds. Every cat has its own "secret" language, known only to its trusted human—and even then, only if that human listens closely enough.

Yet there are still clues that enable a more general system of cat sounds. In this book, I present what I have learned from my past studies and my current research project, "Melody in Human–Cat Communication" (Meowsic). I summarize the various kinds of sounds, the situations in which they occur and the existing variations. I also recount my personal experiences in dealing and communicating with cats. Additionally, the book contains a quick introduction to phonetics, so that my linguistic descriptions can be better understood. Maybe some readers will even give my method a try with their own cats at home. It can absolutely lead to surprises. Or at least to better understanding. It will certainly lead to a better relationship.

Even after years of research, there is still a bit of mystery. But that is exactly why we find our cats so fascinating, isn't it?

1

MY FIRST CATS

Humans and cats: two different species with a common language that bridges the divide between them—is such a thing even possible? Up until now, science has not been able to provide an answer. Yet many cat owners are already persuaded that their own cats can speak. As a cat fancier, I am completely convinced and insist: of course they can speak! But there is also the scientist in me, and she says: I am going to investigate! So it is hardly surprising that I started to examine the thesis "cats have a language" using the scientific method and tools of my discipline, phonetics.

My scientific interest is admittedly directed toward the verbal expressions of cats. Are there "words" that all cats have in common? Can we even call them words? And is there a language that we as humans can understand independently from a cat's other behavior, something that we as humans can study, understand and apply?

But before we start our scientific investigations, let us get to know our five "subjects," the five cats with whom my husband and I share our home: Donna, Rocky, Turbo, Vimsan and Kompis. They are the source of our happiness and the reason for my scientific interest.

I am an early riser. Even if I am still sleepy, I get out of bed and make breakfast for the cats. This morning ritual is the first chance every day to talk to my cats and see how they are doing. Like any other ritual, the breakfast ceremony has a structure.

First, I greet Vimsan, who usually sleeps on the couch in our guest room. While I fill her bowl with food, she hurries toward me, her tail held high, nuzzles and rubs against my legs, jumps onto the sink and mews softly, as though she wants to say "Good morning, It's nice that you are up already. I am hungry."

"Good morning, sweetheart," I say. Most of the time, she leaps for joy, nudges my hand with her head and trills. *Brrrt.* "Thank you."

The triplets—Turbo, Rocky and Donna—are up next. They stand expectantly in front of the kitchen door and greet me with soft trills. Again, *brrrt*, but this time in the sense of "Good morning!" Turbo, our gourmand, who is always up for a treat or a meal, jumps straight onto the counter in the kitchen, trills, purrs and rubs his head against my hand while I prepare his food. I speak softly to all three: "Hi, my darlings. It's great that you are up already. Breakfast will be ready in a second."

Rocky stands on his hind legs, and lifts himself up with his front paws against my knees, where he drawls a *me-aw*, which I take to mean "Oh, that smells good, I want some, too!"

Donna springs gracefully onto a kitchen chair, looks at me expectantly and finally produces an impatient, demanding *mrhrnaaauuu-hi!* Finally, all three of them are in their places and chew eagerly, dedicated to their task.

Kompis has spent the night on his favorite blanket on the footstool in the hall. He stretches and expands to his considerable size, which stands in stark contrast to his bright (with acoustically high resonances) baby meow, *mmeeeheee*. "Don't forget about me, I am hungry, too!" When I put his bowl in its place, he rubs his head against my leg and trills softly, "Thanks!" "You're welcome, my friend," I answer and gently pet his neck.

Then I go out into the garden, where one of the neighbor's cats, Graywhite, resides in her new basket in front of the kitchen window. "Good morning, Graywhite," I say. "Did you sleep well?" When she sees me, she stretches slowly and casually climbs the woodpile with the reasonable expectation that I will put her breakfast on top of it. Graywhite is still very reserved in her behavior with me. I approach her with the necessary caution and try to pet her gently on the forehead. She protests immediately, *Mee, mee!* "No, I do not want that today." "Okay, sorry, I just wanted to say good morning," I say, and go back into the house, where the other cats are waiting for me. The ritual is concluded. All of the cats are satisfied. My day can begin.

The morning ritual with my cats is always interesting. It puts me in a good mood and makes my day more relaxed. Our exchanges, our way of saying "good morning" to one another and of having breakfast together is simply the best way of beginning the day. Even if the procedure always follows the same pattern, the cats continue to surprise me with slight variations. It is always a mixture of friendly and cheerful sounds which vary in their nuances. By now, I can interpret them really well. As a result, I understand my cats better and better.

HOW IT ALL BEGAN

You have surely figured it out already. I am a fan of cats—a *kattatant*, as we say in my language, Swedish. I cannot imagine life without cats. And it has been that way for as long as I can remember.

So I have always looked for, and found, opportunities to get to know cats better, to observe them and to study them. Because I am a phonetician by profession, that is to say, I study the sounds of human speech for a living, I have primarily studied the verbal expressions—the vocalizations—of cats when they interact with other cats as well as with people. The great diversity of different sounds and their nuances is astonishing and differs from cat to cat. The study of this diversity is unending.

And yet there are general patterns in the sounds of almost all the cats I have met. My experience and my discoveries are summarized here and may serve as a kind of phrase book for other cat fanciers. It might help them understand their cats better.

When we understand what our cats are saying better because we are able to listen more precisely, our mutual understanding will be greater. Our relationship to our cats and their relationship to us will be more intense. We will be able to understand and fulfill their needs better and more quickly.

I have loved cats for as long as I can remember. Although we did not have cats at home when I was a child, I asked for one every year, both for Christmas and for my birthday, though I only ever got stuffed animals…

It wasn't until I was an adult that I was able to invite real, living cats into my home. I got my first feline companions from friends and relatives, who either did not want to or could not keep them anymore.

That is how I made the acquaintance of the friendly black-and-white and slightly stiff-legged tomcat Fox, often referred to as "Fox the Hyper" by his previous owner. His nickname was no coincidence. He was always getting agitated about the smallest little things. But as soon as he arrived at my place and left his carrier to probe the terrain of my two-bedroom apartment, he was friendly, gentle and curious. He purred, tried out the food I had put in his new food bowl, made himself comfortable on my bed—and fell asleep.

It was love at first sight, and we lived happily together for many years. When the day that all animal lovers fear came, I had to take a last trip with my old and very sick friend and have him put to sleep. Although I suffered, a life without cats was unthinkable for me. So my husband, Lars, and I took in "vacation cats" and played cat sitters while the cats' owners were away.

Among our favorite guests were the Birman females,

Ludmilla and Estrella, who were as elegant as they were distant, as well as the graceful and highly intelligent gray tabby, Kisseson.

The somewhat fearful but very social beautiful fat black male cat Vincent stayed with us two or three times a year for a few years. Because I liked him so much (and because he hated car rides and carriers so much), we often extended his stays with us in that I returned him to his owners much later than planned.

After a few years, he was finally able to come live with us as our roommate. For seven years, we lived together, loved him, took care of him daily, fed him according to dietary recommendations, and injected him with insulin (he had come down with diabetes) twice a day. The closer he came to his end, the more medicine he needed. At the end we had to give him nine different pills twice a day. He hated it. We needed all of our creativity to convince him to swallow them. The trick that finally did it was the treat afterward.

With Vincent sleeping next to me on his blanket on my desk, I studied linguistics and phonetics, and wrote my doctoral thesis. When he passed away in 2010, we were in despair. I suffered as I had years earlier when Fox passed on. My husband, who is also a great cat fancier, swore, "No more cats. Never again."

Three Cat Kids

"Never again?" Just a few months after Vincent left us, the longing was there again. I started to coax the two neighbor cats—who often passed through our backyard—into our house and fed them treats. I looked at ads for cats that were available for adoption almost every day, and stumbled on a post online about three young black siblings that needed a new home.

They lived in the shed of a nearby community garden, and I was able to convince my husband that we should at least visit them and think about adopting one or two. When we arrived on a cold winter day, the first snowstorm of the season was sweeping through our city. The woman from the local humane society, who had been feeding them every day, had a bad cold. The three little ones were so charming and graceful. We fell for them. But which of the three should we take home with us? And which should we leave in the ice-cold shed? Would we have the heart for it? We tried to simplify the matter by asking the woman from the humane society to decide for us, but she volleyed gracefully: we should take all three home with us, just until they found a home for the others.

The result was already clear on the drive home. "We are keeping all three," my husband told me. That sealed it. The next day, the three cuddly kittens came home with us.

It was the first time we had had such young kittens at home. Soon, my husband and I felt like the parents of young children. There was always something to do. In addition to the normal feeding, litterbox cleaning and vacuuming (three black cats that frolic and play throughout the house lose a lot of black hair), there was always something that the kittens had knocked over or pulled off a shelf.

Even though we were constantly on the move, we did not regret a thing. Donna, Rocky and Turbo (the woman from the humane society had named them already, though they got a lot of pet names from us as time went on), led us on any number of adventures. We were lucky that we always scraped by in the end.

Once on a cold, rainy evening, when we forgot to close a window upstairs, Donna and Turbo somehow managed to climb onto the roof. Rocky wanted to go after them but we caught him just in time. Hours later we captured the two escapees in a rainy nighttime search-and-rescue operation.

Another time we simply could not find Rocky, who is particularly shy. After hours of searching we finally found him hiding in the fireplace. We spent hours coaxing him out, but did not notice at first that he was not just naturally black. Only after he had left soot marks all over the house did we realize the scale of the catastrophe. And that was not the end of our cat adventures.

Three Become Four

A beautiful big red tomcat, who had not been neutered, often passed through our garden. We just called him "Red." His reticence did not stop him from marking our yard as his territory. Logically, anyone whom he identified as a disobedient interloper was chased away. Obviously, he made a great effort to convince the neighbor cats (both neutered females) of his rights. We assumed he had a home. Two years later, we found him injured. A little later, he seemed to be healed. We continued to assume that someone took care of him. Then he seemed to be doing worse again. This time it did not seem like anyone was caring for him.

We packed him up and took him to the vet, but it was too late; the injuries were too serious. Plus, he had developed a tumor. The vet had to put him to sleep. We were in despair. Why had we not seen that he was homeless? Why had we waited so long before we took him to the veterinarian? It was a hard blow for me. I swore to myself that I would never wait again. I would take any cat that seemed to be sick or injured straight to the vet, without wasting time figuring out who the owner was. I had not listened to him, I had not understood him.

A little while later, I built a cat flap for the neighbor's cats, so that they could come warm themselves in our basement in the winter. I filled the small heated room in our basement with food, blankets and water. The next morning, I went to see whether Black-and-White and Graywhite had been to their new sanctuary and discovered their food, but when I went down to the basement, I found a surprise. A totally unfamiliar small gray tabby cat had made herself comfortable on the windowsill and stared at me, her big dark eyes filled with fear and curiosity.

I did not know what to do—I had to get to work. Maybe the cat had just stopped by for a visit. But when I went back to the basement after work, she was still there. I was able to pet her cautiously, and when I did, I discovered a serious large wound on her right hind leg. The whole leg was one large open wound. The fur had been almost totally ripped away and hung on in strips. It was already infected and looked terrible. To the vet! We were lucky to get an appointment early the next morning. The treatment lasted the whole day. Luckily, nothing was broken. The wound could not be stitched; too much fur and skin were gone. One could only hope that the wound would heal itself.

We called the police and put ads looking for the cat's owner in the paper and on the web. Everyone who got in touch went away disappointed. She was not their cat. In the meanwhile, we had given her a name. "Vimsan," Swedish for *bum-wiggler*, because her rear end shook with every step she took, a clear consequence of her injury.

Vimsan is a great cat—but only when she wants to be. She likes to play and cavort with us, but she cannot stand other cats. She likes to lie on our laps and cuddle, but otherwise hates being touched. She never ever wants to be picked up, and if you do anyway, she will bite lickety-split. But we still love this small striped gray-brown cat with the big scar on her leg and the too-short tail (she must have lost the tip in an earlier life).

Four Become Five

Vimsan often got into fights with other cats in our neighborhood. Black-and-White and Graywhite were indubitably among her enemies. When a young black tomcat with white paws and a white chest and belly showed up in our garden sometime in the winter, there was a fight almost every day. The young unneutered tomcat was extremely interested in Vimsan, but she did not want anything to do with him. There were fights high up in our apple tree, in our hedge and on our lawn.

One day, the newcomer showed up with large wounds on his cheeks that just would not heal. He did not seem to have a home. We took him to the vet, cleaned and treated his wounds for several weeks, and looked for his owner. But nobody responded to our ads.

The cat had become a good friend by then. He liked being with us while we worked in the garden or drank a coffee outside. We called him "Kompis" (Swedish for *buddy* or *friend*). And—you guessed it—we kept him. By now his wounds have healed and he has gotten a bit fat, but we love him just the way he is.

Although he is our biggest cat, he has the smallest baby voice, it is even higher in pitch (melody) than the voice of our smallest, Vimsan.

With that, it was decided: our family of five cats is complete with Kompis.

CATS AND PHONETICS

I am a phonetician by profession. I do research and teach at Lund University, in the very south of Sweden. Phonetics, my area of expertise, studies the sounds of human speech. As part of my research, I ask the following questions: How are these sounds produced, and how do they differ from each other (acoustically or auditorily)? I analyze the spectral characteristics (how the sound energy is spread across all frequencies) as well as the prosody (the melody, rhythm and dynamics of speech) in different words, utterances, dialects and languages. Among the hazards of my professions is a tendency to listen more to how something is said than what is being said.

That is how it is with human speech and, as a cat fancier, it is no surprise that I began to listen to the phonetic properties of cat sounds as well. I started to ask myself what vowels were present in a meow. How does the pitch or melody change in the meows of my cats when they are asking me to play with them? And does the melody change in different situations or contexts, such as when they want to be let out into the garden or when they have hidden inside a closet and I accidentally closed the door?

I still remember how I first noticed that my cats meowed differently when they were asking for food at home and when they were in the carrier and on the way to the vet. The melody as well as the vowel sounds of the meows sounded completely different. How could that be? Can it be coincidence? Do cats vary their meows instinctively, or do they learn the different nuances of vowels and melodies and how to use them in different contexts or situations? Could it even be that cats have learned to deliberately use different sounds and their variations in different situations?

At that moment, my love of cats and science first

came together. I started to record the various sounds that Donna, Rocky and Turbo made and analyzed them using phonetic methods—the same ones I normally use when I investigate human speech. Using my "phonetic ears" I listened closely to the sounds, tried to transcribe them using the symbols of the phonetic alphabet and investigated their different phonetic characteristics. In which high and low frequencies could my cats vary their meows? Which cat sounds are voiced and which are voiceless? Which vowels and consonants can cats produce, and how do they move their mouths—their tongues, lips and jaws—when they produce the different sounds?

I read a lot about the different cat sounds, primarily in scientific books and articles. I found that there was remarkably little phonetic research on cat sounds. I took it upon myself to change this.

CAT SOUNDS: AN OVERVIEW

The scientific investigation of cat sounds is, in itself, nothing new. Charles Darwin wrote about cat sounds. He recognized six or seven different vocalization (or sound) types and was especially interested in purring because it is produced during both inhalation and exhalation.

Marvin R. Clark (1895/2016) goes a step further in his book *Pussy and Her Language.* He refers to the work of the French natural scientist Alphonse Leon Grimaldi, who had ascertained that the vowels *a*, *e*, *i*, *o* and *u* can be used to form almost every word in the language of cats and that the liquid consonants *l* and *r* occur in the majority of all utterances. Other consonants, he argued, occur only rarely. If we follow Grimaldi, the language of cats consists of about 600 basic "words," which are used to form all other "words." We also learn from Clark's book that the language of cats bears a strong resemblance to Chinese in that both have only a few words, but those words change meaning depending on pronunciation—especially in relation to the tone (intonation, melody) of the language. Both languages are therefore very pleasant to the ear, almost like music. Modern scientists do not take Grimaldi's book all that seriously, though some of his descriptions can be accurate.

Mildred Moelk published the first (as far as I know) phonetic study of cat sounds in 1944. She listened very carefully to her own cats and organized their sounds into sixteen phonetic patterns divided into three main categories. She also used a phonetic alphabet to transcribe or write down the different sounds; purring, for example, is given as [ˈhrn-rhn-ˈhrn-rhn…] and meowing becomes

[ˈmiɑou:ʔ]. Today, cat sounds are still often divided into the three main classes suggested by Moelk.

1. Sounds produced with a closed mouth, the murmurs (purring, trilling)

2. Sounds produced when the mouth is first opened and then gradually closed (meowing, howling, yowling)

3. Sounds produced with a mouth held tensely open in the same position (growling, snarling, hissing, spitting, chattering, chirping)

Moelk operated on the assumption that the various acoustic patterns in the sounds signaled different messages, for example acknowledgment, bewilderment, request, greeting, demand and complaint.

Jennifer Brown and her colleagues Buchwald, Johnson and Mikolich investigated the sounds of both adult cats and kittens. They found acoustic similarities in various sounds produced in similar behavioral situations, as well as differences in sounds produced in different situations.

Between the 1950s and the 1970s, there were also a number of studies of laboratory cats. Due to the laboratory setting, these studies involved the analysis of unnatural (probably often desperate) sounds that were recorded in the sterile atmosphere of the laboratories where the cats were caged and probably starved before the recording sessions. Luckily, there are now more case studies that were conducted under humane circumstances, often in the private homes of the cats. There are now many scientific studies of cat sounds stemming from behavioral research (ethology) and zoology, as well as an increasing number of linguistic and phonetic studies.

Though there are more recent studies, many descriptions of the sounds of cats continue to refer to Mildred Moelk, her three main categories and her sixteen different sound patterns. I, too, am guided by Moelk in this book and describe most of her sound patterns. On top of those, I also include sounds that were described in other works and sounds which I have recorded and analyzed in my own studies. The categories (sound patterns) are organized according to their phonetic traits or features. Because of the great number of different variations, I have decided to describe only sound patterns that I have personally observed in my own or other cats. I have also recorded the vast majority of these sounds myself and have analyzed them using phonetic methods. I would like to invite you to listen to them yourself and maybe to compare them to the sounds made by your own cat. You will find the relevant links to the individual video and sound examples on my website at http://meowsic.info/catvoc.

In the following pages I will give you a brief overview of the most common cat sounds. A few have two or more names. Books, articles and websites on the topic sometimes use one word and sometimes use another to describe the same sound. Meowing, for example, is also sometimes described as miaowing, and howling is often described as yowling. Because it is likely that these names are used for the same type of sounds, I included the most common name for each sound type first and then included other common terms within brackets. A few examples also include transcriptions using the International Phonetic Alphabet. The individual phonetic symbols are explained in Tables 3, 4 and 5 on pages 260–265. Please do take a look and see if you know some of these sounds from your own cats.

1. Sounds produced with an open mouth

    a. **The Purr**: A very low-pitched, sustained, relatively quiet, regular sound produced during both inhalation and exhalation: [↑hːr̃-↑r̃ːh-↓hːr̃-↑r̃ːh] or [↓hːʀ̃ː-↑ʀ̃ːh-↓hːʀ̃ː-↑ʀ̃ːh]. A cat purrs when it is content, hungry, stressed, in pain, as well as when it gives birth or is dying. Purring probably indicates something more like "I am no threat," "please leave everything as it is," or "keep on doing what you are doing," than "I am content." Mother cats and their young often communicate with purring, probably because it is a quiet sound that is hard for other predators to detect. Purring is also common among some large wild cats—one of the best known is the cheetah named Caine. Many cats can simultaneously purr and trill or meow.

b. **The Trill** (Chirr, Chirrup, Grunt, Murmur, Coo): A relatively short and often soft sound that is frequently rolled softly on the tongue. Trilling sounds somewhere between a purr and a meow, almost like a voiced rolled *r* (although sometimes a little harsh): *mrrrh*, *mmmrrrt* or *brrh*, which can be written in phonetic script as [mr̃ːh], [mːʀ̃ːut] or [bʀ̃ː]. A trill is used during friendly approach and greeting, when playing, and sometimes as an acknowledgment or confirmation (which might be interpreted as "yes, got it" or "thanks"). A trill can be varied in pitch and trilling, and has the following subcategories:

    i. **The Chirrup** (Chirr): A more high-pitched trill, often with a tonal rise at the end of the sound.

ii. **The Grunt** (Murmur): A shorter and more low-pitched trill, often with level or falling tone.

iii. **The Murmur** (Coo): A soft nasal sound without any trilling or rolling *r*-sound, which sounds more like a soft [m] or murmuring sound.

It is not unusual for a trill to turn into a meow, thus producing a more complex sound: *brrriu* [br̃iuw], *brrmiau* [bʀ̃:miau] or *mrrriau* [mhr̃:iauw]. Purring and trilling can also occur together.

2. Sounds made while the mouth is first opening and then closing

a. **The Meow Sounds**: This is one of the sounds used most frequently with us humans. It has a great number of different meanings and phonetic subcategories. Meowing mostly takes place with an opening mouth that then closes. I have identified the following subcategories based on their phonetic characteristics.

i. **The Mew**: A very high-pitched meow, often with the vowels [i], [ɪ] and [e], sometimes followed by a [u], i.e. [me], [wi] or [mɪu]. Kittens often use this sound to get their mother's help or attention. Adult cats may mew when they need the attention or help of their humans.

ii. **The Squeak**: A raspy, nasal, high-pitched and often short mew-like call, often with the vowels [ɛ] or [æ]. A squeak often ends with an open mouth: [wæ], [mɛ] or [ɛu]. Squeaks are often friendly requests for attention.

iii. **The Moan**: A somewhat dark (with acoustically low resonances) meow, often with the vowels [o] or [u]: i.e. [mou] or [wuæu]. A moan is often used by a cat that is either anxious, stressed or demanding something.

iv. **The Meow** (Miaow): A sound which often includes a combination of multiple vowels that often, but not always, produces the characteristic [iau] sequence, i.e. the typical meow sound. Meows are often directed at humans in order to gain their attention, and may sound like [miau], [ɛau] or [wɑːʊ].

v. **The Trill-Meow** (Murmur-Meow): A combination of a trill and a meow sound. Often with a rising tone, which may sound like [mʀ̃hŋau] or [whr̃ːau].

b. **The Howl** (Yowl, Moan, Anger Wail): A long and often repeated sequence of extended vowel sounds, usually produced by gradually opening the mouth wider and closing it again. A howl may consist of a combination of vowels and semivowels, such as [ɪ], [ɨ], [ɤ] or [j], or [aʊ], [ɛʊ], [ɑʊ], [ɔɪ], or [ɑɔ], i.e. [awɔɪɛʊː], [jɪɨɛɑʊw] or [ɪːaʊaʊaʊaʊawawaw] with a rising and falling melody. It is used as a warning signal in aggressive and defensive (agonistic) situations, and is often merged or combined with growling in long sequences with slowly varying melody and loudness.

c. **The Mating Call** (Mating Cry): A long sequence of meow-like sounds, trill-meows and/or howls produced with an opening and then closing mouth by both female and male cats. The sound sometimes sounds a bit like a human child weeping and crying. Perhaps that is why humans often react instantly to this sound.

3. Sounds that are produced with an open tense mouth are often associated with offensive or defensive aggression, but also with sounds directed at prey

   a. **The Growl** (Snarl): A guttural, harsh, very low-pitched, regularly pulse-modulated sound of usually long length (duration) produced with the mouth slightly open during a slow, steady exhalation. A growl often sounds like a very deep and trilling *r*: [gʀː], [ʀː], or a creaky [ɹ̰ː] or [ʌ̰ː]. Growling is used to signal danger or to warn or scare off an enemy, and is often combined, with howling and hissing.

   b. **The Hiss and The Spit** (the more intense variant): A voiceless fricative (noisy) sound often produced with an open mouth with a raised upper lip, visible teeth and an arched tongue, with a hard exhalation. A hiss is often a warning and deterrent sound, but may also be an involuntary reaction to when a cat is surprised by an (apparent) enemy. The cat changes position with a startle and breath is forced rapidly through the slightly open mouth before stopping suddenly: [fːhː], [çː], [ʃː] or [ʂː]. Spitting is more explosive, sometimes with a *k*- or *t*-like sound at the beginning of the sound: i.e. [t͡ʂ̻ː], [k͡hː̮] or [k͡ʃː], and sometimes a little saliva is even expectorated.

c. **The Snarl** (Scream, Cry, Pain Shriek): A very loud, harsh and often high-pitched sound produced just before or during active fighting, often with [a], [æ], [au] or [ɛʊ] vowel qualities. A snarl is sometimes used as a final warning, but injured or sick cats may cry when they are in pain.

d. **The Chirp and The Chatter** (Prey-Directed Sounds): Sounds that are sometimes produced around prey (birds, rodents, insects). A hunting instinct where the cat attempts to imitate the calls of the prey or the killing bite, for example when a bird or an insect catches the attention of the cat. There seem to be several subcategories based on their phonetic characteristics.

    i. **The Chatter** (Cackle, Teeth Chattering): A voiceless, very rapid, stuttering or clicking sequence of sounds produced with the jaws juddering, which produce a crackling *k*-consonant: [k̟̚⁼ k̟̚⁼ k̟̚⁼ k̟̚⁼ k̟̚⁼ k̟̚⁼] or [k k k k k k].

ii. **The Chirp**: A voiced, short call, said to be mimicking the chirp of a bird or rodent. The pitch is often monotone or falls toward the end: [ʔə]. It is generally repeated in sequences [ʔɛʔɛʔɛ]. Softer, weaker variants have also been observed like soft **tweets** without any clear initial [ʔ] and with varying vowel quality, for example [wi] or [ɦɛu]. There are also variations where a soft chirp or tweet is prolonged, so that it almost sounds like a tweedle or warble, and with rapid changes in the pitch or melody. These variations are often combined with tremor or quavering: for example [ʔəɛəɥə].

Now, having already read my descriptions of the most important cat sounds, there is no point in reading any farther, or is there? Well, in the chapters ahead I would like to explain a little bit more about the most important cat sounds and above all else, I would like to discuss some of the situations or contexts where they typically occur.

2

CAT FOR BEGINNERS

There are a whole lot of cats in the world—there are 95.6 million cats kept as pets in the United States, 10 million in Canada, 10 million in the United Kingdom, 3 million in Australia, and 1.5 million in New Zealand as of 2017 (McNamee 2017). Despite there being so many cats, many people cannot properly interpret the sounds their cats make. However, anyone who listens a little bit more carefully when cats "speak" to us will quickly understand that they are able to produce a great number of different sounds, and that it is not very hard at all to learn to interpret the different sounds. An example: although our cat Kompis prefers to spend his time in the garden, he likes to sleep in the house, especially when it is cold out.

His favorite place is a big blanket-covered stool in front of the radiator in the hall. He often sleeps for hours, but when he wakes up and wants to go out we know right away because he uses sounds that lie in the range of frequencies to which we humans are especially receptive. "Meeaahh," he says, in a very high-pitched and bright (with acoustically high resonances) voice. Even if we are upstairs, and very far away, we can still hear him. We also know that Kompis has a much deeper voice when he tries to scare off an intruding cat in the garden. On those occasions it sounds more like a low-pitched *moouuoouu*. How does he know that we understand him better when he uses his bright and high-pitched voice? Why does he change his voice when he interacts with other cats? Can cats learn how to best communicate with different species (and individuals)? Behavioral scientists and biologists have already learned a lot about how cats communicate. Can we phoneticians contribute anything to the understanding of feline communication? The differences between human speech and animal sounds are well-known. Discovering the similarities and building bridges for better understanding is especially interesting for me as a linguist.

But first let us talk about the differences between the language of humans and the language of cats. So as to clarify the difference between them I first will discuss how cats communicate in general terms, and then turn to a detailed description of the range of cat sounds.

CODES OF COMMUNICATION IN HUMANS AND ANIMALS

We humans prefer verbal, that is to say spoken, communication. Although one often hears talk of the "language" of bees, apes, dolphins or whales, a great number of researchers have recognized that their communication cannot really be described as language. Many scientific investigations have confirmed that the vocal (acoustic) codes of all other species are not only simpler, they are also more limited compared to human speech. It is unlikely that future research will discover an animal species whose means of communicating deviates from this pattern. In addition, human language is open, meaning we can add a limitless number of new words with new meanings. Animals, in contrast, communicate about a very limited number of topics. They can discuss "here" and "now," but usually not "yesterday," "next week," "over there" or "in Sweden."

When apes, cats or other animals communicate with sounds, a single sound usually corresponds to a single "word" with a specific message within certain contexts or situations (one that the hearer often interprets as a meaning). The words of human speech, in contrast, are composed of multiple small parts, like the consonants and vowels (phonemes), that contribute to the overall meaning. We can change the meaning by changing one of these parts, such as with *cat* and *bat* or *house* and *mouse.*

Animal sounds depend on their context, and though they may be meaningful, they do not consist of smaller parts that themselves can change the meaning of the sound, such as the consonants and vowels of human speech. If a cat first says "mew" and then says "meow," the two sounds do not necessarily mean different things. A communication code with thousands of different meaningful sounds needs, among other things, a very complex apparatus such as the human voice box with which to produce those sounds—something that simply does not exist in the animal kingdom. Or does it? The most recent research suggests that many animal species do have a kind of "languageness" that is not exactly like human language, but which is not necessarily simpler or less successful as a communicative code.

HOW DO CATS COMMUNICATE?

Cats and humans have lived together for more than ten thousand years. We domesticated them. But they probably domesticated us, too. They taught us how we should best behave around them (do not approach too quickly, do not handle too roughly, do not speak too loudly). We made it clear to them that we were happy to have them around, that we like to feed them and pet them, that they can expect warmth and protection from us, as long as they are just a little friendly to us and occasionally catch a mouse or two, so that our grain stores are not emptied by rodents.

Although many cats are solitary animals who rarely seek the company of other cats, cats can live together in friendly groups. Additionally, most domesticated cats seem to like living with humans. In this sense, they are social creatures who communicate in a variety of different ways with each other as well as with us humans: through scent (olfactory), with body postures and movements (visual), through touching (tactile) and with sounds (acoustic).

Humans, unfortunately, are not hound dogs; we are not especially sensitive to scent or the pheromones that cats can detect so easily. Moreover, our eyes are often occupied by watching our smartphones, computers, books, newspapers, magazines, televisions and so forth, so we might not notice that Kitty has been sitting next to her empty food bowl waiting for breakfast for more than half an hour. Perhaps that is why cats and humans have developed a kind of acoustic language that both species are able to understand. Cats have understood that sometimes, the best and quickest way to get what they want from us is to communicate with sounds, a *meow*, for example. They know that we will react immediately and we mostly know what our cats want from us: for us to give them food, open a door, retrieve a favorite toy mouse from under the sofa or just spend half an hour of our time petting, cuddling or playing with them.

Touch: Tactile Communication

Our cats know very well that the best way to communicate with their humans is with sound. Even so, they have maintained other forms of communication. Nose-touching, head-bumping, and rubbing against us, head to head, head to body, or body to body, and kneading our laps with their paws (we call the rhythmic stepping of a kitten against the teats of its mother and of a grown cat against a soft surface such as a blanket *kneading, kneading dough*, or *making biscuits*). Sometimes they will also show us they have had enough with either their claws or a bite. All of these are examples of tactile communication. Touch is very important, not only between mother cats and their kittens, but also between cats belonging to the same social group. It is possible that cats want to use touch to show us that they accept us humans as their friends as well.

Cats that have befriended each other like to lie close to one another when they rest or sleep. Moreover, they may groom each other. Head- or cheek-rubbing with another cat, a dog or a human is often also a form of greeting. This tactile communication consists of friendly, affiliative gestures and serves to reinforce social cohesion.

Body Postures and Movement: Visual Communication

We should pay far more attention to the visual signals of our friends. Postures and movements, either of the entire body or of individual body parts, such as the tail, head, face, and above all, ears, eyes, and whiskers, provide clear indications as to the momentary mood or needs of the cat. In aggressive or defensive situations, the increase in the volume of the body by arching the back and raising the hair (usually referred to as piloerection) mostly means that the cat feels threatened and is meant to signal "I am big, and I have muscle and sharp teeth which I will use, unless…" However, it might often just be a bluff. That is why cats, before they attack, often assume an upright posture (often with an arched back), bristle their fur and sometimes open their mouths, so that they seem as big and scary as possible. This is how they defend themselves from any potential attackers.

On the other hand, when Vimsan crouches flat rather than standing or sitting in an upright posture whenever Donna passes by her, this usually means something very different: she is signaling that she is very small and harmless and does not want to provoke a fight under any circumstances. Cats frequently communicate with subtle visual signs, for example, through their head, ear and eye postures and movements. There are also less subtle signals, like moving the body and the tail, as well as raising their fur, but these symbols also subside quickly. By no means do they last as long as a scent marking, for example.

Slow movements like the closing of eyes, yawning, cleaning or even creeping away in slow motion demonstrate peacefulness and harmlessness. Rapid movements, in contrast, (tail wagging, foot stamping, and running toward or away from an enemy) are mostly signs of excitement. They indicate that it can get serious at any moment and that a fight might be in the making.

Tail signals are especially interesting. A tail held vertically often means "I am young, small and friendly." A tail held upright, but puffed up like a brush can often mean "I am big and impressive." In contrast, a tail held upright with a slight kink or hook resembling a question mark often means "I am contented, curious and friendly."

Tail wagging has a fundamentally different meaning with cats than with dogs.

It rarely has to do with joy or with joyous excitement. Instead, it seems to be more like a reflexive reaction to an inner conflict. The stronger the wag, the stronger the conflict. While a slow wagging is often only a sign of intense concentration, stronger wagging means "I am excited," and even stronger wagging means "I am very worked up—it is about to get serious."

Spraying and Rubbing: Scent Signals

Unfortunately, we humans cannot perceive all of the scents that our cats leave behind. The scents called pheromones, which are essential in communication between cats, persist longer than sounds and even continue to communicate something long after the cat that left this scent message behind has gone elsewhere. Scent markings are almost like writing for cats. These signals can describe the cat's gender, age, health and readiness to mate, as well as reveal how old the scent mark itself is. Scent marks wear off with time and need to be continually renewed. Urine, stool and scratch marks are all among the scent signals. Scratching or rubbing with the head or the body also leaves decisive scent marks behind, as cats have scent glands on their paw pads (between the toes), as well as on their head and cheeks.

We humans often entirely misunderstand these signals. When cats spray urine around or scratch the furniture, we often take it poorly and suspect malice. And we do quite a bit to prevent this behavior in our cats.

We try to clean our furniture or treat it with chemicals that we hope will force our cats to give it a wide berth in the future. In the worst case, we might have to dispose of furniture that has been so treated. For the cats though, this behavior is communicative and important for the care of their claws.

For example, when Kompis pees on the bushes in the garden, he tells all other cats that the garden is his home territory. In cities, where there are many four-legged creatures in a tight space, nobody gets their own beat, but instead must negotiate the same territory.

Amazingly, cats in tight quarters are often able to reach compromises and establish a kind of shift operation: "I can patrol here in the mornings without a problem and leave my scent marks all over the place, the neighbor cats can come in the afternoon (when I am at home sleeping anyway) and do the same. That way, we rarely meet each other and avoid conflicts." That is how Kompis solves the problem. His warning to other cats—that he is the king of "his" garden, young, healthy and ready to defend his territory—is understood by potential rivals. If another cat does come and leave a mark on his territory, he renews his claim the next day by leaving a fresh mark.

When Turbo sharpens his claws on his favorite scratching tree, it does not serve only to help maintain his claws;

it is also a scent marking. The glands on his paws leave a scent on the tree, so that the other cats can tell that he was there. It is a kind of cat social media—"I am logged on, and this scent is my status update."

Rocky and Donna also communicate with scent when they rub their heads against my legs, my face, the kitchen door or the leg of a chair. These marks mean something more like "I live here, and I feel well. I would like to leave my scent here so that the residents and things all smell like me. That way, I feel safe and secure."

Even if I cannot entirely perceive these scents, I have noticed that it smells a bit like bananas when my cats rub their foreheads or cheeks against my face. I do not know whether I am just imagining it, but to me, this scent says something like "You are my human, and so we should both wear the same perfume." Maybe it is a kind of declaration of love or at least a way of ensuring the togetherness and belonging of cat and human.

TIP: If you have a sufficient number of scratching posts and trees in carefully selected spots (where the cat is feeling secure and comfortable) you may reduce the scratching of your furniture.

Meowing, Trilling, Growling and Purring: Communication through Sounds

Not all cats communicate happily or frequently with sounds. Many prefer to be silent. Let us not forget that they are predators, and being a predator is deeply embedded, even in our pet cats. That is why they instinctively try to hide their location and their physical and mental state (especially when they are sick, in pain or giving birth to kittens) from other animals or humans to avoid conflicts. Still, they sometimes want to communicate with each other with the help of sounds. Cats like to be out and about at night, and sounds are an especially sensible way of communicating over long distances or when visibility is bad. Certainly many are familiar with the nocturnal concerts of cat sounds.

Cats have learned that humans react well to cat sounds. We humans do not have the same good noses that cats have, and our eyes are also often elsewhere, so that we do not notice, for example, when our furry friends have sneaked into the kitchen and sit in front of their food bowl. When we are working, are occupied with our computers or smart phones or are sleeping, sounds are especially effective. The four-legged companions have understood that and adapted themselves to us. That is why many cats develop a kind of spoken language together with their humans that is mutually understandable.

I have also found that the more I talk to a cat, the more it talks back to me. However, it is important to clarify something here: Do all cats speak the same "language"? Can they understand each other when they communicate with vocal signals? There seem to be signals that are universal and are understood by all cats. But there also seem to be geographic, cultural and breed-based differences. Maybe cats are even influenced by the language or the accent of the people around them. When I give lectures about cat communication, people often come to me afterward with comments and questions. For example, "My cats make entirely different sounds than the ones you played in your lecture. Could it be because I usually speak Japanese at home with them?" Although it has not yet been investigated thoroughly, many researchers are of the opinion that cats can, in fact, develop family, group or neighborhood dialects (Bradshaw, 2013; Leyhausen, 2005). Do cats have dialects or do they develop a set of unique sounds that only their humans can understand? This fascinating question is also the subject of my academic work.

Now we are getting down to it: I would like to explain my work as a phonetician briefly and then help you understand the sounds of cats from that perspective.

WHAT DOES A PHONETICIAN ACTUALLY DO?

My task as a scientist is primarily the investigation of human speech. I have been doing it since 2000. It sounds simple, but some knowledge of the methods is necessary.

My natural curiosity makes my work a lot easier. When I am tracking down a secret, it is not so easy to scare me off the trail. Even if everything is smooth and flawless on the surface, I like to scratch a little bit to see if something else is not hidden there after all.

How do we produce spoken language? How are the sounds of speech (vowels and consonants) produced in different languages and dialects? What do they sound like? These questions are at the core of my work and continue to catch my interest. I am also interested in how sounds, syllables, words, phrases and utterances vary in length (duration), tone (intonation, melody), loudness and voice quality—that is to say, in prosody, as well as what they sound like in different languages and dialects. I have also observed the changes our human voices undergo as we grow older using scientific methods. And it gets even more interesting: How do our emotions influence the way we speak? Why does our speaking style change depending on with whom we speak? Why do we sound different when we speak to young children than we do when we speak to elderly people, when we speak to someone we love and to someone we dislike, or when we speak in private and professional capacities?

We change the intonation or melody of our speech even when we pronounce a simple word, *cat*, as a question or as a statement. When we say "Cat." as a statement the intonation generally falls, whereas when we say "Cat?" as a question the intonation rises. The sounds also arrive at our ears in very different versions. For instance, vowels are frequently louder than consonants. In the word *kittens*, the vowel *i* is the loudest sound. We emphasize it much more heavily than the other sounds, including the vowel *e*, which in casual speech often is not pronounced at all, although we use it when we write the word. The four consonants sound very different. Try it yourself! Pronounce the letters *k*, *t*, *n* and *s* one at a time, and listen very closely. You will be able to hear that the *k*, which is produced with the tongue farther back than the *t*, sounds darker (with acoustically low resonances). It becomes even more apparent when you compare the pronunciation of *n* and *s*, where the *s* sounds much brighter than the *n*. These differences can be explained in that the main sound energy of *n* can be found on lower frequencies, while *s* has most of its sound energy in the high frequency bands, and *t* usually has sound energy on higher frequencies than *k*. Moreover, did you notice that the *i* and the *n* are voiced, as they are produced with vibrating vocal folds (vocal cords), while the *k*, the *t* and the *s* are voiceless? Such phonetic characteristics, as well as a number of others, can be further studied in visual representations depicting the acoustics of speech. At the end of this book starting on page 259, you will find a phonetic alphabet containing all the vowels and consonants I use to describe the sounds of cats in this book.

The following figure shows a three-part diagram which is commonly used by phoneticians. The upper pane shows the waveform (often called an oscillogram), which is a representation of the microphone signal in a recording of me saying the words "Kittens. Kittens?" with my Swedish accent. (I pronounced the first word as a statement, the second as a question.) In the waveform, we can see how loud and how long the different speech sounds are.

In the middle, you can see a spectrogram—it shows how the sound energy of each speech sound is distributed across different frequencies. Because vowels are generally pronounced louder than consonants, they typically also have more energy, and so they show up darker (blacker) in the spectrogram. The *s* is dark in the upper range of frequencies, but completely white in the bottom range. That means that this *s* has no energy in the deeper frequencies. Instead, its energy is entirely concentrated in the higher range of frequencies. In an *n*, exactly the opposite is true—lots of sound energy at the lower frequencies of the spectrogram, but none at all at the top.

In the bottom pane of the diagram, the fundamental frequency (the acoustic term for the pitch contour or melody) of the words is tracked, that is to say, how our tone of voice rises and falls when we speak. You will see right away that the melody of "kittens." (statement) and "kittens?" (question) is different.

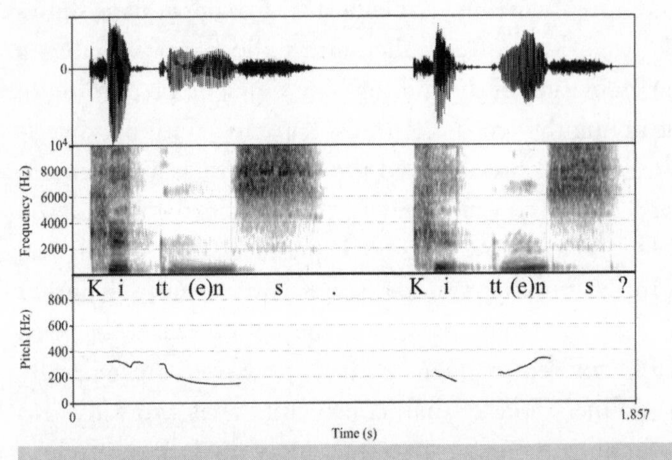

Three phonetic diagrams for the word "kittens." (statement) and "kittens?" (question): Waveform (top), spectrogram (middle) and fundamental frequency (pitch, melody) (bottom).

Furthermore, we have determined that the same speech sounds are pronounced differently in different dialects and languages. For example, I used a method called electromagnetic articulography, which can be used to track the movements of our speech organs, to determine how vowels are pronounced in different Swedish dialects. I literally looked inside the mouths of different speakers in order to see how they move their tongues, jaws and lips when they pronounce different vowels.

I also translated these vowels into phonetic writing. To aid me I had a system that works in every language: the International Phonetic Alphabet (see Tables 3, 4 and 5 with the phonetic symbols at the end of this book, pages 260–265). Phonetic transcription depicts sounds as they are pronounced. One symbol per sound is the rule. My pronunciation of the word *kittens*, for example, can be transcribed [ˈkɪt(ə)n̩s].

If these phonetic methods work for every human spoken language, I said to myself, they might also work for cat sounds. And, as I have discovered, they usually do.

One of the most commonly used methods of my academic discipline is acoustic analysis. With the help of a computer we can measure different acoustic features of the sounds of speech and compare them. We can measure the length of a sound, such as an *e*, in milliseconds, and we can measure the intensity (loudness or volume) in decibels. Moreover, we can determine the frequency distribution of a sound signal (a speech sound or a word) when it is visually depicted in the form of a spectrogram, just like the one I have provided above.

In a spectrogram we can see, for example, that the sound energy of an *e* is distributed across entirely different frequencies than those present in an *a*, and that an *m* mostly has energy in a lower spectrum of frequencies, while an *s*, on the other hand, is mostly concentrated in the higher range. The fundamental frequency, that is to say the part that we normally perceive as the pitch or melody of speech, can be measured using acoustic methods. We can measure precisely how high or how low the pitch of an individual speaker is, whether a phrase or a sentence has a monotone melody or has tonal highs and lows, whether the melody rises or falls or maybe does both. Acoustic analysis is objective, which means that the results are always the same regardless of who conducts the measurements (at least as long as they enjoyed a basic education in phonetics).

The way the human ear perceives sound depends on the individual listener. It is subjective. A number of factors influence the way that we hear—age, experience and hearing loss, for example. For that reason, many researchers in my field conduct perception or listening tests. In such experiments, a group of listeners are asked to listen to sounds, words or sentences that are pronounced by speakers who speak either in the same or in different dialects. They then compare word A and word B, for example. Are the words pronounced in the same dialect, with the same intonation, the same melody? How old is the speaker? Are sound 1 and sound 2 the same or different? The results from all participants are then compiled and the averages are used to show how the sound stimuli were perceived by the majority of the listeners.

Phoneticians also concern themselves with categorization within linguistic systems, describing the number and type of vowels, consonants, melodic patterns and other traits that characterize a dialect or a language. In this subfield of phonetics—phonology—the rules that govern the combinations of sounds and syllables in a language are studied. English, for example, allows for combinations of consonants at the beginning of a word, such as in *stripe*, which is not possible in many other languages (e.g. Japanese and Finnish).

This knowledge, by the way, is a precondition of teaching computers to speak (speech synthesis), and related to teaching them to understand human speech (automatic speech recognition). In speech technology, phoneticians work alongside engineers to develop synthetic speech that can be used, for example, in GPS systems for cars.

These are all subfields of the phonetic sciences. I was trained in them during my studies and have applied their methods in several research projects. Still, it took quite a while for me to come up with the idea of applying them to cat sounds. I started by perceiving the many sounds my cat made using my "phonetic ears," and made recordings of them. The more closely I listened to the sounds that came from Vincent's mouth, the more questions I had—questions that could only be answered using the methods of phonetics.

A MEOW IS NOT A MEOW IS NOT A MEOW: CAT SOUNDS FROM A SCIENTIFIC PERSPECTIVE

What does Turbo actually want when he meows in that particularly strange way? And what kind of funny chattering noises does the normally shy, even fearful, Rocky make when he sits on the windowsill and sees a bird in the garden? Do these sounds have any meaning at all? Why do cats purr anyway, and what exactly does it mean? When and why do cats use sounds to communicate with humans? Why do we (humans and cats) seem to understand each other so well? Or is it just an illusion when we believe that we can communicate with our animals so well through speech and sounds?

Is there even something like Cat-language, Feline or Cat, and if so, how similar is this "language" to our human languages? Could it be that we might understand the sounds our cats make if we look for universal features like pitch, length, volume (acoustic sound level pressure or intensity) and melody rather than for words, grammar, vowels and consonants?

In 2010, I attended a lecture about cat purring held by

my colleague Dr. Robert Eklund in Lund. He opened my eyes (and ears) to the fact that I, too, could apply my phonetic and linguistic knowledge, my experience and my research methodologies, to the study of cat sounds. Not just to purring, but to all cat sounds. I started eavesdropping on my own cats and other cats as soon as I returned home. I made as many recordings as I could so that I could study their phonetic characteristics.

Since then, I have been recording my own and other cats on an almost daily basis. My first big discoveries were that the frequency range in cat sounds is enormous; that cats use a very large number of different sounds in order to communicate with us humans and with their fellow felines; and that the fundamental frequency can rise very rapidly from approximately 25 hertz (number of oscillations per second, Hz) to over 1100 Hz.

All of the sounds of both human and cat language (our vocalizations) are produced when a stream of air encounters an obstacle either in the larynx (voice box) or farther upward in the vocal tract. This may occur in the larynx when, for example, producing a vowel as the airstream from the lungs encounters resistance in the form of the vocal folds held closely together and thus causing them to vibrate, but also at the front part of the palate (alveolar ridge) directly behind the teeth, when we use our tongue to produce a *t* or an *s*, or between the lower lip and the teeth, when we pronounce an *f*. That an *s* or an *i* sounds brighter than an *h* or an *o* also has to do with articulation. The size and shape of the mouth or oral cavity determine the color (or resonance) of a sound.

The organs cats use to produce sounds are smaller than ours, but they are similar. Both humans and cats have larynxes, tongues, palates, lips and jaws. Cats move them in much the same way we do when they produce their sounds. That is why we can also study the sounds of cats using phonetic methods.

Why do cats need so many sounds, and why do they vary their voices so much? What do the acoustic signals (that is to say the sounds) of cats actually mean? And what can a linguist, who is neither a zoologist nor a behavioral scientist, contribute to our understanding?

It is not so easy to study the articulation of cats. One has to be present whenever the cat feels like "saying" something. A cat never vocalizes on command. Not even if you ask nicely, or try to persuade them to say something.

Laboratory experiments are out of the question. First of all, I do not want to put any cat in an unnatural setting. Moreover, you would get only false results, because the conditions under which the results were obtained would be artificial. That means no X-rays, and no electromagnetic articulography, during which small sensors are attached to the tongue in order to investigate the movements of the vocalizing organs. Nor would I even think of using other invasive methods, which could cause the cats unease or even pain. I would never stick thin needles (or electrodes) into the muscles of the organs that cats use to produce sounds. I would never stick a probe into a cat's mouth in order to record a video of the movements of their tongues or vocal folds (vocal cords). Nor would I force a cat to spend minutes tied up in an MRI scanner, which produces very loud sounds. Ultrasound is considered by some to be an option. Even, however, if some animal lovers, veterinarians and scientists are of the opinion that it represents a viable option, there is still the question of how we would convince a cat to purr, meow or trill when their fur has been shaved off, their necks have been coated with gel and a hard device is pressed up against their larynx or under their chin. For that reason I prefer just videotaping the movements of cats' mouths and examining the movements visible to the naked eye, at least until some other truly workable method is discovered.

Video and audio recordings seem like the better option when you want to study cat sounds. You just have to be there when the cat happens to feel like saying something, then hold the microphone close enough to its mouth that the recorded sound is audible and has a quality suitable for scientific analysis. For the most part, cats do not vocalize particularly loudly. What is more, they often seem to be able to tell when one is planning an experiment and become disgruntled. They are liable to scamper away, hide and probably will not make a peep. The research would be much easier if one could simply hold the microphone up and politely ask the cat to share a little by meowing three times, purring twice and, as a crowning conclusion, hiss once or twice, and then explain to me what they usually mean when they utter these sounds. But failure is preordained with that method, so I decided on a different strategy. I prepare my "research area" by arranging the camera, microphone and treats. Then I hope that the cats whose actions I want to record will perceive the environment as completely normal and so will produce natural sounds, expressing themselves exactly as they always do when it is time to be fed, when they want to be let in or out, when they want contact with their humans, to cuddle or play. I try to keep as far in the background as possible, so as not to influence their behavior and expressions.

I listen to the recordings I have made this way a number of times using my "phonetic ears." I describe the sounds using phonetic terms and transcribe them with the phonetic alphabet.

After that, it is time for the acoustic analysis. I study the distribution of frequencies of the sound energy in spectrograms. I measure the length, fundamental frequency of the melody, the intensity and other characteristics.

Using my results, I can try to categorize every single sound according to a system. To which sound category does a particular vocalization belong? What does the melodic pattern look like? What cat (gender, age, breed) made the sound?

Only then can I begin to find potential answers to specific questions. Do cats who are excited or stressed have different meow-melodies than relaxed cats? Do all cats make similar sounds in the same situation? Which variations can be found within a sound (sub-)category? These questions can be addressed very well using phonetic methods. In the upcoming pages, I will present a research project in which I investigated the melody in the communication between cats and humans. (See Chapter 12, Studies and Projects, page 204.)

In the following chapters, I will describe the most common cat sounds in more detail. I will depict every type of sound and its phonetic subcategories and will also present the situations in which each sound usually occurs. Additionally, I will address the related body language (postures and movements), articulation (position or movement of the mouth) and phonetic transcription, as well as the voice and melody. And I will present concrete examples of every sound type (with vocal expressions from my own or other cats).

Milch

3

"MEOW"
THE MOST COMMON WORD IN THE LANGUAGE OF CATS

In the vocal (acoustic) repertoire of the cat, there is nothing as common as a meow. The meow seems to be the preferred sound when cats address humans. Its sound can be varied and nuanced almost infinitely. Some cats meow with a dark (with acoustically low resonances) voice, while others use a bright (with acoustically high resonances) voice. Many cats meow frequently, others hardly meow at all. Still, every human knows exactly what a meow sounds like: meowing is, without question, *the* cat sound. It runs the gamut from the volume of a fire alarm to scarcely audible, and there are even completely silent meows. Our old cat Vincent's penetrating meow was on the alarm end of the scale. When he was alive, we did not need an alarm. He knew exactly what to do when he was hungry in the morning—he just needed to meow repeatedly to get his humans out of bed so that they would make him breakfast. His wake-up call worked for a while, but we humans learned quickly, too. In order to stop being "trained" like that, we stopped serving breakfast immediately after we got out of bed. And it seems to work, as we are very rarely woken by meowing now. With five cats it would be a nightmare if we were! Although I make breakfast for my cats early in the morning, it is never the first thing I do when I get out of bed. I always find some time to read the news, check my email or go for a walk before I feed them.

No matter their stage of life, each of our cats likes to meow in his or her own special way. Kompis meows with a bright "baby voice" when he wants to be let in or out. Rocky meows in two or three syllables when he wants to play with us or with his siblings. Turbo meows with a hoarse voice when he absolutely has to sit on my lap. Donna meows softly and coaxingly when she tries to get me to play or cuddle with her. Little Vimsan seldom meows, but has learned slowly that a soft mew helps when we do not immediately understand what she wants (e.g. to be fed or to be let out into the garden). If that does not help either, she gets clearer. She meows more loudly, incisively—and effectively. She is similarly loud when she is stuck outside and wants in, as though she understands that she needs to turn up the volume if she wants the people inside to hear her.

DESCRIPTION OF THE SOUND

My cats have shown me that there are many different meows. The sound exists in countless variations and is used in very different circumstances. A meow can be made with different vowels (e.g. [iu], [iau], [uæ]), with or without a closed-mouth *m* (with an *m*: [miau], without an *m*: [au] or [waʊ]). A meow can also have different numbers of syllables ([wu-au], [mia-wau], [miæ-æ-aʊ]).

A meow can be assertive, coaxing, demanding, inviting, imperious, whining, melancholic, suffering, friendly, brave or undaunted. It is often used to get attention ("I want something") or to make a declaration ("my bowl is empty again"), but it can also simply be a friendly greeting ("I see that you are there. I am here"). There may even be a meow that we humans cannot hear—a kind of ultrasonic meow with frequencies that are not perceptible to us humans. (Cats can hear much more of the ultrasonic spectrum than we can, and can therefore also hear the noises of their prey, such as mice, very clearly. They may even be able to produce ultrasound noises.)

As a rule, meows are produced with an opening-closing mouth. This is how the typical meow sound is produced or articulated. The *m* is produced with a closed mouth, the mouth then opens for the *e* and stays open for the *o*, before closing with the *w*. Try meowing yourself, and look in the mirror while you do it. Do you see how your mouth first opens and then closes?

Some meows start with an open mouth *aou*. Others end with an open mouth *wuea*. Although *meow* is written very similarly in a great number of different languages and often begins with the letter *m*, the sound actually often begins with a [w] or a [u]. Cats also sometimes meow in two or more syllables (such as *meowow*). Female cats, who have smaller vocal organs, often have brighter meows than tomcats, while kittens meow more brightly yet.

A meow can be varied almost without end, and because there are so many different versions, it is not simple to assign the different sounds to subcategories. Different individuals have different types and nuances in their meows—perhaps because every cat adapts its meow sounds to the particular situation and needs, as well as adapts them to its humans. There may even be geographical differences—meaning that there may be differences caused by the influence of human speech (dialects) that are spoken around the cat, as well as by the sounds made by the other cats in the area. Differences between breeds also play a role. For example, many people claim that Siamese cats have a particularly loud meow. Some breeds are supposedly more talkative than others, and some cats communicate much less than their siblings from the same litter. Many can meow polysyllabically, for example [mi-a-a-au] or [wa-æh-æh]. It is almost as though they are speaking in sentences, with multiple words. Maybe that is because they have listened to us speaking in longer sentences and want to try it out themselves.

Every cat owner has to listen to their cat patiently and with good ears in order to start to understand the meaning of particular sounds. But the phonetic characteristics introduced in this book will set you on the right path. I have chosen the following categories of meow based on their phonetic distinctions, so as to depict the great variation in cat expressions.

1. The Mew

A mew is a very bright and high-pitched meow, which often contains [i], [ɪ], [e] and [u] vowels. The mouth is sometimes open very little, sometimes slightly more. This sound is often used by young cats when they want the attention or help of their mother. Kittens often mew when they are cold, hungry or lost. We must assume that mother cats both perceive and understand this sound very well—experience shows that the mother cat comes quickly when the kitten mews. There are adult cats who continue to use this baby language with us humans when they are in distress or afraid.

2. The Squeak

What we hear when a cat squeaks is similar to a mew, but raspier, hoarser, more nasal and often shorter. The vowels that are present are often [ɛ] or [æ]. This friendly, inviting sound usually ends with an open mouth and sounds like [wæ], [mɛ] or [ɛʊ]. My cats often use this sound when they are trying to get me to play with them or give them a snack. Experience shows that squeaking is usually used to say "I want something and I am glad that you see that," or "I am so small and cute, I want something from you." The melody often rises at the end.

3. The Moan

A moan is a darker, often plaintive or woeful sound, frequently containing the vowels [o] or [u]. Cats often moan when they are uncomfortable, anxious, nervous or stressed—for example, when they are locked up in a room or a cat carrier and cannot get out, or when they absolutely have to have something and are very demanding. It sounds like [mou] or [wuæu]. The melody is frequently (but not always) level and declines at the end.

4. The Meow

The typical meow sound is produced using a combination of vowel sounds to produce the characteristic sequence, which can be written in the phonetic alphabet as [miau], [ɛau] or [wɑːʊ]. However, a meow may also include other vowels. Meowing, along with trill-meowing, is the most common sound in the communication between cat and human. It is used to get our attention, for example when we are standing in the kitchen preparing something delicious, or when we are supposed to be made aware of an obstacle, such as a closed door or window. We as humans are often very sensitive to this sound and react immediately. Kittens, as soon as they are no longer babies, meow predominantly at their mothers. Adult cats meow primarily to humans and only rarely to other cats. The meowing of adult cats may be understood as a consequence of domestication. We can evaluate it as the continuation of the mewing of kittens to humans.

5. The Trill-Meow

When a meow is introduced by a trill, a complex sound arises: the trill-meow. This sound is also very common; because it begins with a closed mouth it may belong to a subcategory of the meow. It sounds like [mr̃hiau], [mhʀ̃ŋ-au] or [whr̃ːau]. A deeper tone (intonation, melody) is typical at the beginning of the sound, though the pitch (melody) quickly rises as it becomes a meow. Trill-meowing is very common when a cat is vying for our attention and is one of the most common sounds used with humans.

CONCRETE EXAMPLES

As mentioned before, my cats do not all meow the same amount or as frequently as each other. Furthermore, all five have their own special meow variations. Vimsan often mews or moans softly. She only recently started to meow "properly." After we found her injured, she had to wear a cone collar so that she could not lick her wound.

We felt so sorry for her as she mewed softly. We did not even know if she was in pain or just didn't like the collar. She continues to mew or to moan when she is hungry or wants attention. We think that her positive experience with mewing (we helped her when she was injured) is still fresh in her memory and that she uses it now when she begs for sympathy and wants our help. Once, she hid in the attic of our house. We did not notice and closed the door. After a few hours, we heard a soft mewing from above, which we identified as the typical Vimsan mew, though we were not able to locate it at first. It got louder and louder until it led us straight to her and we were able to free Vimsan from her disagreeable situation.

In contrast, Donna is the queen of squeaking. She does it when she wants to play or to cuddle. She relies on getting her way, and not without reason. She is a spoiled princess, and she has her retinue well under control. Squeaking is her method. She can vary the strength of the signal so effectively that we, her caregivers, immediately understand what she wants and needs.

When we bring our cats to the vet, there is always a loud moaning in the carrier. Except when Rocky is the passenger. He is so afraid in that situation that he hardly says a thing. The strange environment, the foreign sounds and smells—no cat likes them, and mine certainly do not. They moan without ceasing. The tone always falls toward the end. It is heartbreakingly sad, like a child's lament to its parents. Additionally, Kompis suffers from claustrophobia. His moaning and howling sound especially desperate. He also scratches and clamors particularly fiercely in an attempt to escape from his temporary prison. As good cat parents, we have a strategy for this unavoidable situation. We always try to get the first appointment of the day at the vet's so that we can take him out of the box as soon as we have been shown into the examination room.

Donna, Rocky and Turbo often meow when I prepare something tasty for them in the kitchen. They are especially eager for fish. If the gang of cats notices that fish is on the menu, the meowing picks up immediately, and in a very demanding tone: "Gimme!", "I want some, too!", "Another piece, please!" Though all three meow especially at me, there is something unique about Rocky—he also meows at his siblings when he tries to get them to play. He walks around the house meowing with an ascending melody, *meow*, sometimes with two syllables *weowow*. "Where are you? Don't you feel like playing?" Turbo uses an especially long and hoarse meow when he thinks that I have neglected him and wants to cuddle or play with me.

Our friends Peter and Marie have a beautiful male cat named Zoran. He figured out quickly that the best way to get his people to open the basement door is to sit in front of it and meow loudly. He is successful most of the time, and then he can run into the basement, where it is quiet and cool.

The combination of trilling and meowing is frequently a friendly greeting combined with an invitation to play, cuddle or give treats. Among our cats, Donna is the one who has perfected this sound. If squeaking or trilling does not help, maybe because I am too occupied with my computer, she comes and explains to me with a loud trill-meow that I should darn well stop what I am doing (usually a senseless occupation in her view) and should dedicate myself to whatever she thinks is important at the moment. Right now, the most important thing for her is her favorite toy, a fishing rod with bright feathers. It only makes sense that the invitation to play is connected to it. For her, playing is a priority of course, and she has no patience for procrastination. "Come play with me," she seems to say.

Turbo and Rocky also produce trill-meow sounds when we get home from work and they want to say hi. We also take the sounds to mean that they would like something to eat, a little play or a cuddle, though they are not as unrelenting in their demands.

TIP: Now, if you want to listen to the sounds that I have described only theoretically and learn what they actually sound like, you could turn to the appendix on pages 237–251, where I have collected a number of links to sound and video examples of different cat sounds. These examples are available on the website http://meowsic.info/catvoc and can be found under each sound category described there. As I am learning more about the different sounds of the cat almost every day, the project is still in progress, and the website is regularly updated with new examples of sounds and nuances. Therefore, you may find more examples on the website than are described in this book. I would like to show you the large phonetic variation within every sound category. I hope you will enjoy listening to them, and perhaps even recognize some of the sounds.

Body Language

Body postures or movement of either the whole body or individual body parts—as well as the small facial expressions that accompany a meow—are always dependent on the particular situation. When producing attention-getting meows ("I am hungry," "I want to go out," "I want you to stop working and play with me," "I do not like it here, I want in/out/away") cats often seek eye contact with their humans and often express their desires with their body language. They put themselves near the situation or the object that they want. They will stand in front of the door, for example, when they want to be let out, or in front of their empty bowl when they want to be fed.

Occasionally, they will even tilt their heads to one side and look at their humans with big eyes, as though they were begging. Therefore it is generally easy for humans to understand what a cat wants. Many of them stroke or rub their heads or their entire bodies against the legs of their humans when they meow in order to make the behavior that many humans designate as begging clearer yet.

PHONETIC CATEGORIZATION (SOUND TYPE, MELODY)

Articulation

Most meow sounds are made with an opening-closing mouth, but there are also variants that either begin or end with an open mouth. Mewing is sometimes even created with a nearly or entirely closed mouth, a kind of bright, singing murmur or coo, *mmmmm*, almost like a trill, but without the rolling. We do not know exactly if and how the tongue moves because, as already explained, we are not able to observe it using current research methods.

Phonetic Description and Transcription

Meow sounds consist primarily of a combination of two to three vowels, though single vowels are also sometimes present. The vowels in question are often [i], [ɪ], [e], [ɛ], [æ], [a], [ɑ], [o], [u], [iu], [ɑːʊ], [ɛʊ], [æu], [oɑu] or [iau]. Sometimes consonants are present at the beginning, and less commonly at the end—[miau], [ɛaw] or [wɑːʊ].

TIP: The phonetic symbols are described in the tables starting on page 259.

Voice and Melody

Meow sounds are voiced and have a rising or falling melody. Sometimes a melody that first rises and then falls has been observed. A meow can be monotone, but it can also have great variations in tone, encompassing frequencies ranging from 50 to 1000 Hz, according to the situation or context and the mood of the cat.

The great variation in the vowels and melody is probably dependent on the situation and depends on how urgently the cat wants something or on how emotionally charged the message is. Most cat owners learn to interpret the nuances of their cat's meows very well. A scared cat sounds different than a happy or annoyed cat. A cat who wants something urgently uses a different, often more varied melody ("I want it *now*") than a cat whose desires are not as urgent.

Meowing is often encountered along with the following sounds: trill (trill-meow as a friendly greeting or invitation), meowing in two or more syllables, meo-o-ow, wow-ow and purring (if the cat is content). In the following, we will dedicate ourselves to trilling and its variants.

4

"BRRRH, HOW NICE TO SEE YOU!" GREETINGS AND SMALL TALK

There is hardly anything nicer than being greeted by a gentle trill when you get home. It sounds so friendly and sweet that you feel that your roommate must have missed you a little after all. Maybe Kitty is not quite as independent as everyone says.

When our three black-furred siblings were young, Turbo had a bad habit. He crept into our room at night, crawled under our bed, hooked his claws into the bed frame and swung around happily like a monkey for a while. He trilled with abandon the whole time. Of course he always woke us up. The pleasure was definitely one-sided and we decided to close the bedroom door at night, although we otherwise liked having our roommates nearby. The late nights had an upside though. I learned to recognize Turbo's trilling voice. I can now tell which of my cats is vocalizing by their voices alone (without looking), and even in the dark. In the meantime, Turbo has quit his climbing exercises and the door of the bedroom remains open.

Our cats' voices are just as distinct as those of humans. Donna's trilling is much brighter (with acoustically high resonances), more like a high, rolled *r.* Rocky and Kompis trill in higher voices than Turbo. Vimsan only trills very brightly, gently and quietly. Do you have more than one cat? If you take the time and make the effort, you will doubtlessly be able to identify and recognize the individual voice of each cat.

DESCRIPTION OF THE SOUND

Most people assume that purring is the friendly cat sound, but in my experience trilling is even more friendly, as it signals happiness, affiliation and friendship, while purring may sometimes signal stress or even distress. Trilling often occurs in combination with meowing (trill-meowing as a friendly greeting or demand for human attention) and purring when they feel good and want to be petted.

Trills (also called chirrups, chirrs, grunts, murmurs, or coos) are rather short and usually soft sounds often produced by rolling the tongue. They almost sound like a voiced rolled *r*, and are sometimes a little rough, something like *brrh*, *prrret* or *mmmrrrt*, though it is likely that they are produced farther back in the mouth.

Mother cats greet their kittens with a soft trill when they return to the nest and they use a similar sound—a bright trill with a rising intonation—when they ask their young to follow them. A trill between mother and her young is used in greetings and to coax. It is no wonder that cats who live with humans trill and chirrup in their interactions with humans, sometimes imitating the relationship with their mother and behaving as though they were young. They often use a short, soft chirrup, grunt, murmur or coo to signal recognition when they encounter their humans or a friendly cat and want to say hi. They also use this sound to nicely ask their humans for attention or if they want to show them something and want to ask their humans to follow them.

The friendly, attention-getting chirrup sounds are often pronounced with a rising intonation, maybe because a question or a request is implied, for example when asking for something to eat or for a human to play or cuddle with them. There is also certainly a little anticipation of happiness in the sound, as most cats know that their wish will be fulfilled as soon as they have gotten the attention of their human. That may be why chirruping and trilling sound so friendly and cheerful.

The deeper grunt is used more as a greeting or as a confirmation than as a means of getting attention. This sound may sometimes sound a bit like a threatening low-pitched growl, but it is much shorter.

The more high-pitched and brighter chirrup, which often rises in tone (intonation, melody), is more striking and is easy to interpret as it generally counts as friendly.

For many cats, an attention-getting trill produced with a closed mouth turns into an open-mouthed meow. A *brrh* can become a *brrmeow*, or a *mrrreeooww*, for example. This complex sound—the trill-meow—is often a friendly sign of impatience, when the cat thinks that their human should finally get with the program. Purring and trilling can also occur together. That is why I have concluded that trilling is the friendly cat sound par excellence and that the animals are happy and contented when they use this sound around us.

CONCRETE EXAMPLES

My cats often trill, chirrup and grunt, and I understand these to be friendly sounds. We hear trills and chirrups of greeting especially frequently. With five cats at home, it can be a regular concert of trills when my husband and I get home from work: "Hi, there you are! It is nice to see you again, have you already noticed that my bowl is empty?"

A high-pitched or bright chirrup (often with a rising tone and sometimes followed directly by a meow) can mean "I want to play," "I want to go out," "I want to be petted or fondled," or "I want something to eat."

Even our big muscular tom Kompis often greets me with a high-pitched, soft chirrup when he is outside and sees me standing at the kitchen window. He knows very well that I understand him to mean "I would like a treat, please," and will open the window right away to give him something to eat.

An even brighter, more high-pitched chirrup, with an even more ascending melody, frequently combined with a meow, can mean (depending on the cat's mood or needs) that something is really urgent. "Please, I am really very hungry now," "Please let me out right away or I will pee on your carpet," "Please, I want you to stop playing with your computer and pay attention to me right now!" Donna knows very well how best to get me to play with her. She starts by trilling softly and continues to trill louder and louder until I will definitely hear her. If I still do not do what she wants, she starts to trill-meow with an even louder, more high-pitched, brighter tone. Finally, I cannot stand it anymore, admit defeat and play with her for a few minutes.

Cats in heat also trill, murmur or coo softly. Before Vimsan was spayed, she mewed and trilled softly when she was in heat.

Some trilling (the grunt) is rather low-pitched, deep, and sometimes sounds a little hoarse or raspy. Turbo often greets his siblings, Donna and Rocky, or me with a low-pitched, short grunt. At night, he sometimes makes an even more low-pitched grunt when he sneaks into our bedroom at night: "I am awake, why are you sleeping?" Both Rocky and Turbo use the deep trilling or grunting as a kind of confirmation, as though they want to say thank you or "good that you have understood me," for example when I give them their breakfast in the morning.

When Turbo sleeps in his basket on my desk and I wake him with a gentle stroke, I often get a short, satisfied trill in response. I take it to be a sign of satisfaction: "Yeah, I am here, and everything is good."

Donna has a large repertoire of sounds when she would like to cuddle. She sits on my lap and sings in squeaking, trilling and purring tones. It is music to my ears, and an exceptional sign of friendliness. I cannot possibly resist it; I always give in and cuddle her, forgetting the work I have to do at my computer in the process.

TIP: You will find a list with links to examples of the various trilling sounds at the end of the book. See the appendix starting on page 237.

CORRESPONDING BODY LANGUAGE

When friendly cats meet, they often greet each other not only by trilling, but also with a little nose kiss, that is to say they bump their noses or foreheads together.

Another gesture commonly used in greeting consists of one cat softly poking its head into the side of the other and then proceeding to nuzzle the length of the other cat. Anal inspection—bum sniffing—also belongs to the gestures associated with trilling. As we humans are significantly larger than our cats, the only interactions of these that take place with us are rubbing their cheek, body or tail against our legs. As I have already described, they leave behind scent signals on our pant legs, skirts or socks in the process. It is a kind of scent stamp marking us as "their human."

Trilling can take place while sitting, standing or while moving. Kompis often sits in front of the kitchen window and trills, as though to say "I would like a treat now." Our Donna can trill happily while she runs like lightning through the house, simply because she has noticed that I have stood up from my desk and am following her to the toy basket.

Turbo can even trill while he sleeps in his basket. It's as though the spot triggers this reaction in him. It seems more important to him than body language or movement.

I have observed my cats very carefully in the situations in which they trill in order to better understand this sound. Low-pitched grunting often occurs when approaching, as a sound of friendly greeting and as a confirmation (as though they were saying thank you). High-pitched chirruping sometimes functions, in contrast, as a request for attention. "Please get up and get me something to eat."

PHONETIC CATEGORIZATION (SOUND TYPE, MELODY)

Articulation

Trilling is probably always produced with a closed mouth. The vocal folds (vocal cords) vibrate in the process, but we still know very little about the position of the tongue. Because the mouth is closed, the air escapes through the nose; it is a nasal sound.

Phonetic Description and Transcription

Trilling sounds are voiced, usually nasal sounds that often resemble an apical *r* or a rolled rear guttural *r*. A coo, murmur or a mixture of a mew and a trill also occurs from time to time—a long [m:] without trilling. Among the typical phonetic transcriptions are [mr̃ːh], [mːr̃ːt], [m̰:] and [bʀ̃ː]. Trill-meow sounds are complex, such as [mhʀ̃iaʊw] or [br̃ːiau]. The squiggle above the *r*, called a tilde, means that the air is expelled through the nose and not through the mouth.

> **TIP:** The phonetic symbols are described in the tables starting on page 259.

Voice and Melody

Trilling sounds are very soft, voiced sounds. The more high-pitched chirrup often has a rising melody, the more low-pitched grunting, and the murmuring or cooing, is usually rather monotone, but there are exceptions. A trill can also have a declining melody, or a melody that first rises and then declines. The frequency is around 350 Hz (a little higher, around 600 Hz, for trill-meows), while the entire spectrum of frequencies ranges from around 100 to about 1000 Hz.

5

"GRRRRRR, HSSSSHH, GET AWAY!"
ANTIPATHY, REJECTION, DETERRENCE

Picture this: it is four o'clock in the morning. My husband and I are sound asleep. Suddenly, I hear a terrible sound, almost like a small child in horrible pain, crying out heartrendingly for their mother. After the initial shock passes, it becomes clear—it is just Kompis, explaining to a rival outside in the garden that there is nothing to see here, and that the interloper should very well be on his way, or else. It is the same drama every spring! This time, the interloper does not surrender so quickly—they give as good as they get, and so the howling gives way to growling. They are locked in a howling, growling duel, with no end in sight. After a while, the interloper admits defeat and creeps dejectedly away. The victorious Kompis has defended his kingdom and begins to clean himself, licking his imaginary wounds.

I assume many of you are familiar with this kind of situation, too. I have observed a great number of similar occurrences in my neighborhood (when I go walking or jogging in the morning, for example), and I have managed to record some growling and howling sounds with my video camera. Frequently, two cats howl together as though in a duet. The dominant voice leads the melody up and down, and the other voice accompanies with weaker, brighter (with acoustically high resonances) tones. It is not just male cats who moan at their adversaries like this. Females, too, can howl and growl at each other for several minutes if they do not like each other. Cats rarely engage in physical violence with each other; they seem to be able to defuse tensions through these concertos of howls, a kind of diplomacy before things go bad. Frequently (though not always), the winner is the one who can howl the deepest, loudest and longest. There are different reasons for this, and some of them are anatomical. An animal with a larger body also has a larger apparatus for making sounds—larger lungs, voice box and vocal folds (vocal cords). So big dominant cats can produce the deepest, loudest tones. Or the other way around: cats howling in the most low-pitched, loud voices appear to be big and dangerous (although they may very well be bluffing). The loser cowers and creeps ever so slowly away, as though in slow motion.

Sometimes, however, there is simply no option but a physical confrontation. When that happens, the howling culminates in terrible shrilling and very loud snarling, shrieking and crying sounds. It is enough to freeze your blood. Luckily, this kind of confrontation usually does not last long.

DESCRIPTION OF THE SOUNDS

The aggressive and defensive (agonistic) cat sounds belong to varying phonetic categories, and are almost all produced with the mouth tense and more or less open. Howling constitutes an exception, as it is produced by first opening, then closing the mouth. There are numerous subcategories, both voiceless (hissing, spitting) and voiced (growling, howling/yowling, snarling/crying/screaming/shrieking). They can sound very different, although they are all used in situations where the cat feels threatened or a fight is at hand. I have defined the relevant phonetic categories as follows:

The Growl

Growling (sometimes called snarling or gnarling) is easily recognizable through its guttural, raw tonal qualities, as well as through its regular, pulsing rhythm. A growl is a very low-pitched and deep, extended sound that is produced as air is slowly and stably expelled through a scarcely open mouth. It sounds like *grrr*, or like a very deep and gnarly vowel [ʌ̰ː] or an *r*-like sound, [ɹ̰ː]. Occasionally, a grumbly [m̰] begins the sound, so that it sounds like *mrrr*. Sometimes this sound is also called snarling, and there is an even deeper, rougher and stronger (louder) version of growling which may be called gnarling or grumbling. All growling sounds signal danger, and they appear threatening in order to warn or scare off the opponent. Growling is often combined or merged with howling in long sequences where the pitch (melody) and loudness slowly rise and fall.

The Howl or Yowl

Howling or yowling is a long and frequently repeated vocal (acoustic) warning or threat signal, which sometimes sounds like *aaaooooouuuu* or *yyyyyooooouuuw*. These sounds are usually produced with a slowly first opening and then closing mouth. The melody can repeatedly rise and fall in the process, and the sound is sometimes short, but often very long. Howling is typically shorter than yowling. Especially when two cats meet at one's garden watch post, long concerts of howls and yowls are not unusual. In a threatening situation these sounds are often combined with growling in long sequences, with slowly rising and falling melody and loudness.

One variant of howling is growl-howling or howl-growling, in which a growl repeatedly transforms into a howl during a rising melody and then back into a growl as the melody falls. The consonants of the growl and the vowel of the howl are combined in long sequences, such as [gɹ̰ːawɪjɑoʀː].

The Hiss

Hissing is a warning signal that can hardly be misunderstood. The upper lip is often raised to reveal the teeth, the tongue is arched and a hard puff of air is expelled. The result is a sharp *hssshhh* or *fffhhh*, which clearly means "Enough!", "Do not come any closer or I will attack!" It is similar to the sound produced by an aggressive snake. Maybe all cats are instinctively afraid of snakes and adapted their sound as a means of scaring away adversaries. However, not only aggressive or angry cats hiss. Cats who become surprised or are insecure also use it as a warning signal.

One can hear a mother cat hiss when she wants to tell her kittens to stop doing something, or when she warns them of danger. Hissing can be an involuntary reaction produced when a cat is surprised by a threat, either real or apparent. The cat then quickly changes position, exhaling quickly in the process. The air is expelled forcefully through a narrow, scarcely open mouth. This produces a sound like [fːhː] or [çː], which then usually stops abruptly.

The Spit

Spitting is an intensification of hissing: a strong exhalation where air is expelled forcefully through a scarcely opened mouth: *kshhht!* [kʃːt]. It is a powerful, intense and hostile-sounding noise that resembles the sound made by a spitting human, *tshhh!* The cat sometimes stomps the ground with her front paws in the process and sometimes also expels a little saliva. It is not only domesticated cats who spit. Wild cats such as cheetahs also spit. Robert Eklund investigated these sounds in wild cats using examples recorded on video and concluded that cheetahs also stomp the ground with their front paws and expel saliva. As the owner of domestic cats, I am amazed by the many similarities between the big wild cats and my little "house tigers." Hissing and spitting can occur simultaneously or sequentially in similar dangerous situations.

The Snarl or Cry

Snarling, crying, shrieking or screaming often sounds like a short, bright, very loud and often rough or hoarse scream. These sounds are often produced before or during a physical attack—out of rage, or as the final warning intended to scare or chase the opponent away. Cats who are tortured or injured and in great pain can also be heard to snarl or shriek. A female cat can also be heard to cry in pain when the male removes his penis from the vagina at the end of the coupling process.

The most terrible sound that I have ever heard was when our cat Vincent was very old and sick. His bladder had to be emptied by the vet, as he could no longer urinate on his own. This desperate cry of pain still gives me nightmares today. The sound energy of snarling or crying lies in frequency bands to which we humans are very sensitive because our babies cry and scream at the same frequencies. That is why we are so often woken up by the howling and snarling of cats who battle over backyards.

CONCRETE EXAMPLES

Before Vimsan came to live with us, I had hardly ever heard howling or growling from our cats. Donna occasionally hissed at her brothers if they got too close, but otherwise we heard only friendly or attention-seeking sounds. But when we wanted to introduce our black beloveds to their new roommate, the small tabby Vimsan, they all started to growl, howl and hiss at each other.

For eight days, I followed Vimsan through our house and recorded her—mainly aggressive—interactions with Donna, Rocky and Turbo. Rocky and Turbo mostly avoided the newcomer or lay on a table or shelf and growled while they watched her investigate her new home. In contrast, Donna was not at all interested in a new relationship and followed Vimsan around, howling, growling, hissing and spitting. Luckily, this aggressive behavior began to subside after eight to ten days, leading to a ceasefire of sorts. Many of my examples of aggressive sounds come from this period.

Vimsan demonstrated her command of aggressive sounds impressively when Kompis arrived in our garden as a young unfixed tomcat. Kompis followed her around our garden and climbed the apple tree after her. He seemed very insulted to only ever be greeted by howls, growls and spits. Sometimes there was also a short fight, though Vimsan snarled and shrieked so loudly and shrilly that Kompis in the end gave up his attempts to get closer.

TIP: Listen for yourself and be impressed! At the end of the book you'll find links to samples of growling, hissing, howling and snarling in the appendix, starting on page 237.

CORRESPONDING BODY LANGUAGE

I have often observed two rival cats in our own or a neighbor's garden, where the cats are sitting down a couple of meters apart staring at each other. They often sit like that without moving for a long time, and when they do occasionally move, it is usually in slow motion. The cat who seems to be at a disadvantage looks as though he'd really like to make a quick exit, but he seldom does. Cats seem to know that if they don't move very slowly, but try to run away quickly, they might be seen as prey. They would run the risk of being hunted and captured, or at least of becoming the subject of a serious physical assault.

In such situations, the ears of the cat are often laid back, the tail whips back and forth, and the fur is puffed up. If the cats cannot prevent a brawl with their sounds, then the worst happens. During the fight, the cats (or sometimes just one of them) snarl and shriek in order to frighten their opponent.

The distance between the cats seems to be important, as does uninterrupted eye contact. My experience and studies have shown that the sounds employed depend on the distance between cats, as depicted in the following diagram.

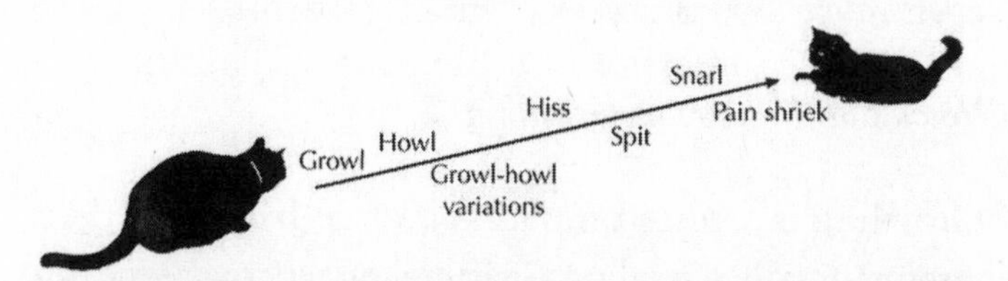

The relationship between the sounds normally uttered by cats in an aggressive or defensive situation and the distance between the cats.

PHONETIC CATEGORIZATION (TYPE OF SOUND, MELODY)

Articulation of Growling

Growling is produced using a slow and steady exhalation. The lips are stiff, tense and slightly open, as is the mouth. The vocal folds vibrate slowly, producing a very low-pitched and regularly pulsating sound.

Phonetic Description of Growling

Growling almost sounds like a very low-pitched sustained voiced trill. We might transcribe it as [ɡʀː] with a vocalic rhotic [ʀː] or as a really low-pitched, creaky vowel [ʌ̰ː] or rhotic (*r*-like sound) [ɹ̰ː]. Occasionally, growling begins with a creaky [m̰], as in *mrrr...* Gnarling or grumbling is sometimes described as a rawer, deeper and louder and even more low-pitched variant of growling.

Voice and Melody of Growling

Growling is a voiced sound that often has a raw, deep, rasping or trilled quality. Gnarling is deeper yet—next to purring it is the deepest sound in the cat's repertoire. The melody has a fundamental frequency (pitch contour or melody) of between 70 and 100 Hz, though falsetto growls at around 200 Hz have also been observed. Growling and gnarling are often interspersed with howling.

Articulation of Howling

Howling is a prolonged vocalic sound normally produced with an opening-closing mouth. When two rival cats meet, they often howl a duet; one seems to follow the melody of the other. Especially when they meet on the backyard beat of one of the cats, a long howling concert is not uncommon. In threatening situations these sounds are often combined with growling in long dynamic sequences, with a repeated increase and decrease of the pitch and loudness.

Phonetic Description and Transcription of Howling

Howling consists of a combination of vowels and semivowels, such as [ɪ], [ɨ], [j.], [j] or [ɤ]. The sound produced is something like [awɔɪɛʊː], [jɪɨɛɑʊw] or [ɪːaʊaʊaʊaʊawawaw]. Dipthongs such as [aʊ], [ɛʊ], [ɑʊ], [ɔɪ] and [ɑɔ] are also very common.

Howling and growling are frequently interspersed in long sequences. In the process the pitch and loudness rise and fall repeatedly and slowly. Cats howl in the same range of frequency as human babies cry, which is why human adults are so sensitive to it.

Voice and Melody in Howling

Howling is voiced and can be very loud, though the volume (acoustic sound level pressure or intensity) goes up and down. The pitch rises and falls in repeated, often irregular patterns and can be of different lengths (durations), though it is frequently quite long. So the melody goes up and down between approximately 100 and 900 Hz. Sometimes a howl is shorter than a second, but longer yowls of up to ten seconds are also not uncommon.

Articulation of Hissing/Spitting

Hissing is produced using a narrow passage inside the mouth, which produces a turbulent noise or fricative sound, when the air is expelled from the lungs. Hissing is probably produced by narrowing the passage between the tongue and the roof of the mouth (palate), so that a rushing sound, also called a fricative, is produced when the airstream from the lungs is expelled. Depending on where the passage is located, the sound can be either dark (with acoustically low resonances) and dull, resembling a powerful *hhhh* (which occurs when the passage is at the back of the mouth). It can also be brighter and more hissing, like a *sh*, such as in *shoe* [ʃː]/[ʂː] or an English careful and noisy pronunciation of the initial *h* in Hugh (or the German *ch*, such as in *Milch* [milk]) [çː] (the narrow passage is farther forward in the mouth).

Spitting is more explosive and can have a short consonant sound [k] or [t] at the beginning of the sound, like in [t͡ʂ ː] or [k͡ʃː].

Phonetic Description and Transcription of Hissing and Spitting

Hissing is not a combination of vowels and/or consonants. Instead, it often consists of a single voiceless sound. Frequently, this is either a dark fricative produced at the back of the mouth, such as [hː], or a bright hissing sound produced at the front of the throat, such as [çː], [ʃː] or [ʂː]. Occasionally, hissing may also start with a sound similar to *f*, such as [fːhː]. Spitting is more explosive and powerful and can begin with a brief plosive or stop consonant, such as [k] or [t], though it quickly gives way to a hissing, rushing or fricative sound: [t͡sː].

It is a typical warning sound, and we humans may—if we are very careful—use it with our cats when we want to warn them of dangers or to prevent undesirable behavior.

TIP: The phonetic symbols used in the text are described in more detail in the tables at the end of the book. (Pages 259-265.)

Voice and Melody in Hissing and Spitting

As hissing and spitting are unvoiced sounds, they do not have a melody.

Articulation in Snarling, Screaming and Shrieking

Snarling, screaming and shrieking are typically produced with a wide-open mouth and sound like a short, bright, loud and often raw or hoarse scream. Just as with howling, snarling, screaming and shrieking often have their sound energies in the same frequency range as the screams and cries of a human baby, so we react strongly to these sounds.

Phonetic Description and Transcription of Snarling, Screaming and Shrieking

Snarling, screaming (or crying) and shrieking typically consist of short vowel sounds (as far as I know, consonants are rare), usually [a] or [æ], though dipthongs such as [aʊ] or [ɛʊ] are also sometimes present in these sounds. If there is even a difference between snarling and screaming or shrieking it probably lies in the tone of voice. Snarls are darker and more low-pitched than screams and shrieks, which are often quite high-pitched. Moreover, there may be a difference between snarls and shrieks on one hand, and screams and cries on the other; it probably consists of the length of the sound. Intense pain often causes long screams, whereas snarls or shrieks usually are shorter.

Voice and Melody in Snarling, Screaming and Shrieking

Snarling and shrieking are short, loud, voiced and often hoarse or rough sounds. The melody is often monotone (often around 300–550 Hz) and sometimes declines slightly at the end. Screaming can be longer in duration, and often has a higher fundamental frequency.

6

"MEEEOOOWWW, I WANT YOU!" CALLING, COURTSHIP, SEDUCTION

Like humans, cats can be burdened by the desire for physical love. Once, Vimsan rubbed up against my legs, looked at me imploringly and unhappily, then rolled back and forth across my feet, trilling and meowing a few times before looking longingly out the window. I said, "Sorry, my love, I am not letting you out. You have not been spayed yet and we do not want any kittens around here." Vimsan sang the whole night through. She combined loud meowing with rather plaintive singing and soft trilling. *What luck that we have an appointment at the vet next week*, I thought.

By now, Vimsan has been spayed and does not make

such sounds. For about a week, I recorded video of Vimsan in heat, including her other sounds (e.g. soft mewing, trilling, cooing), but forgot to record her loud, desirous meowing—her mating call or cry—at night.

What is a mating call, actually? Is it only female cats who call, or do males do it as well? Up until now, I have not collected any evidence to indicate that there is a special category of sound that cats use only when they are trying to attract a sexual partner. Most people know the typical traits of these sounds, which ring out every spring. However, if we pay closer attention, we will quickly realize that the calls of seduction and the sounds used to frighten others away can be very similar.

Maybe it is just a louder, more melodic and sustained meowing when love is at stake, whereas it is a plaintive howling when territory is being defended or an opponent is being chased away. Further studies will clarify the issue.

One can certainly say that it's usually female cats in heat who use mating calls to try to attract a future sexual partner. Unneutered male cats don't go into heat. Instead, they react to the scents and sounds of females who are in heat.

DESCRIPTION OF THE SOUND

Mating calls or mating cries are long lamenting sequences that consist of meow, trill-meow and howl sounds. They are often produced with an opening-closing mouth. There is a lot of variation—some sounds are short, others are longer. What they have in common is that they are repeated over a long period of time—sometimes hours. Some suggest that it almost sounds like a baby crying, or like a small child whimpering for their mother. Perhaps that is why we humans react so strongly when we hear this sound. We are startled and think that we must help our darlings immediately—it sounds so wretched!

But sexual desire can also be signaled with other sounds. Cats who are in heat trill, purr and emit soft meows when they rub up against us, bump their head against us, or lift up and turn their rear ends toward us. A further variant is the loud lamenting calling that can persist for hours or days. The need to call seems to be especially profound at night. Many people are woken on a spring morning by cats who are calling for a mate either at home or in the garden.

As I said above, male cats do not go into heat, they simply respond to the strong scent and sound signals of females who are in heat. The tomcats set out on their trail and can then also call back with meows and other similar sounds. Furthermore, combative sounds like howling, growling and snarling or screaming are also among them, as other males are also enticed and the rivals must spar among themselves to figure out who gets to mate with the female. It can therefore be difficult to determine whether one is hearing "real" mating calls or whether it is simply defensive and aggressive (agonistic) sounds.

Tomcats who have been neutered (especially if they were older when the operation took place) can also respond to the signals of females and sing as if to say "I hear you, I am here and I am ready for you." When a male cat clears up territorial issues with sound signals, it often sounds more like howling.

Mating calls are used to attract a partner, and the frequency code, that is to say the specific highs and lows of the sound, indicate that it is intended to be friendly and inviting, as the melody often rises at the end (see the heading "Additional Phonetic Characteristics: Prosody" in Chapter 9, page 156). As the frequency code is universal, that is to say that it is valid for all mammals, we humans often do the same—when we ask a friendly question the melody of our speech also rises.

CONCRETE EXAMPLES

In our family we have heard the sounds of love from only Vimsan so far. Since Donna is one of three cats from the same litter, neither we nor the vet wanted to risk one of her brothers getting her pregnant (we had noticed them sniffing around her bottom with increasing regularity), so the triplets were fixed as soon as they were old enough.

When Vimsan came to live with us, we didn't know how old she was. Moreover, she was seriously injured and her large wound was infected. We decided to wait for her wounds to heal and let her go into heat once before we had her spayed—much to her displeasure, as she had to stay indoors at the time. For a week, I recorded her trilling and soft mewing, as well as other signs of her heat, such as her coaxing, escape attempts, rolling around on the floor and such. Unfortunately, I do not have any recordings of her nocturnal concertos of sounds. To compensate, I included similar examples from other websites—these can be heard on the website in the category "Mating Call (And Other Sounds by Cats in the Mating Season)" under the keywords "Female cat mating call 1" and "Female cat mating call 2."

The beautiful neighborhood tomcat Red seemed to always be stalking females. A couple of times I filmed him wandering through our garden and recorded some nice sounds as he marked our fence and plants. Many of his sounds are very similar to meows; sometimes he also produces trill-meows.

When Kompis, the big guy among our cats, first came to us, he was still young, thin and virile. Because he had sustained serious wounds on his cheeks, which, to make matters worse, produced an allergic reaction which prevented the wounds from healing, he had to be treated with antibiotics, cortisone and daily cleansing of his wounds for a long time. Only when his infection had truly healed did we get the vet's go-ahead to have him neutered. In the meantime, it was not a pretty picture: the young skinny tomcat with large wounds on his cheeks paced back and forth in our garden, calling for our females. And what a voice! He did not care about his injuries. He seemed completely dedicated to enticing one or both of our females with "cat concerts" that lasted for hours. Longing trilling soon transitioned to expectant trill-meowing and plaintive meowing whenever he encountered Donna or Vimsan in the garden. The poor lad did not know that both our females had been fixed already. And even if his wooing never led to an actual encounter with Vimsan or Donna, the desire and calling persisted.

TIP: In the appendix you will find some links to examples of the beautiful calls of love-crazed cats.

CORRESPONDING BODY LANGUAGE

It is easy to recognize a female cat in heat on the basis of her body language and of her vocalizations. Her natural instinct drives her—she wants to mate and lets the world know about it through

- Continuously seeking attention—rubbing against furniture and other objects, against fellow pets and against humans (especially against their legs or feet);

- Regularly pacing back and forth; frequently rolling on the floor; and

- Assuming the mating position (when she is petted by her humans, for example)—lifting her rear end, sticking her tail up and padding restlessly in place.

Some females also mark walls and doors with a vaginal secretion or a pungent urine that contains estrogen and is supposed to attract tomcats. They also frequently lick their genitals, which are usually swollen when in heat. Cats who are confined to the house will attempt to escape and—if they do not succeed—will scratch the windowsills or drapes in frustration.

Unneutered toms who recognize these signals will generally become restless and answer with regular urination and with aggressive territorial patrolling behavior. They frequently wander around restlessly, marking every bush and fence (as well as a few cars), meowing and calling incessantly. Moreover, they also attempt to track down the female who is in heat. Often, they are not the only ones to have noticed the signals of the female. The competition does not sleep. Rivalries, fights, screaming and howling are preprogrammed. To the victor goes the princess, as well as the whole territory.

PHONETIC CATEGORIZATION (SOUND TYPE, MELODY)

Mating calls or mating cries have many similarities to meowing and to howling, but some typical phonetic characteristics can be distinguished as well.

Articulation

Just like meowing and howling, mating calls (or mating cries) are produced with a repeatedly opening-closing mouth. The articulation (position or movement of the mouth) of mating calls is often somewhat slower than meowing and howling, and is sometimes merged and mixed with extended trilling and somewhat longer vowel sounds.

Phonetic Description and Transcription

Mating calls or mating cries consist primarily of long

emphasized vowel sounds such as [aː], [uː], [a͡ʊ], [ɔ͡aʊ] and [ɪːa͡uː]. They are often preceded by a [w] or a trilled nasal consonant such as [ʀ̃ː] or [r̃ː], producing a sound like [wa͡ːuw] or [rːɪːa͡ʊː]. Trill-meows such as [mhr̃ːwaːoːuːɪː] or [ʀ̃ːwːuːaːu] are especially prolonged.

Voice and Melody

Mating calls (or mating cries) are voiced and are produced with a loud voice in long penetrating sequences. The melody varies, but often rises at the end. Soft trilling, mewing and cooing have also been observed in cats who are in heat. Mating calls can take place over multiple hours at night.

7

"HRRRHRRRR, I AM HAPPY WITH YOU!"
HAPPINESS AND SATISFACTION

Is there a more calming sound than the purring of a cat? Hardly. If you are sad, nothing is more comforting than petting a purring cat. It relaxes a person, makes us happy and peaceful. I still remember the first purring from each of my cats: Fox, who investigated his new home with loud purring, or Vincent, who when I went to visit his previous owners, came into the guest room, hopped up onto my bed and purred for hours.

I discovered that cats have very different personalities when the triplets came to live with us. The small but very courageous Donna came and sat on my lap and purred loudly and confidently already on her first day with us.

Turbo waited to see how his sister would behave. Once he saw that nothing bad happened to her, he took heart and came to me as well. As I petted him softly, he started to purr, too.

In contrast, Rocky was (and still is) very fearful and shy. He spent the first week in his new home in the fabric tunnel that we had purchased as a toy for the cats, emerging only when he was hungry or had to use the kitty litter. After a few weeks, he decided that we were no threat and made friends with us, and he started to feel more at home in our house. He blossomed into our number one cuddle cat. He purrs more frequently and louder than all the rest, and sometimes, when I come into a room where he is lying on a blanket, I can hear before I open the door that he is purring away to himself without any reason at all. Cats really do sometimes purr when they are alone.

The injured Vimsan purred as soon as I petted her for the first time, and although her purring is still very quiet, I often hear a soft buzzing, almost like regular, weak creaking sounds—but when you put your ear up to her body, you hear that she is really purring. Kompis purred at me for the first time when I fed him (he was probably really hungry, and he did not care at all that it was some strange woman, whom he had never seen before, who was giving him something to eat!). Even today, he purrs regularly and loudly when he follows me inside from the garden to get his breakfast or when he is lying on his favorite footstool or my husband's lap and is being petted.

The first recordings of cat sounds that I ever recorded and investigated using phonetic methods were of our old cat Vincent. In the video, he is lying on a blanket on the couch purring. Therefore, I have a special relationship to the sound of purring.

DESCRIPTION OF THE SOUND

Purring is a very low-pitched (often between 20 and 30 Hz), sustained, relatively quiet, fairly regular, humming sound. Cats often produce it for minutes at a time while breathing in as well as out. Most people know this typical cat sound very well and know that it is a sign of satisfaction. But cats do not purr only when they are happy or satisfied. They also purr when they are hungry, stressed, afraid or in pain—even when they are dying. Females purr while they give birth. Purring probably means something more like "I am no threat," or "keep doing that," than "I am happy." It makes perfect sense.

Some say that the sound of a purring cat may have a

healing quality for people. Both lower blood pressure and an antidepressant effect have been ascribed to it. Moreover, purring may be good for cats. It seems that it releases endorphins that help cats to calm themselves. However, to my knowledge, there are no scientific publications supporting these claims. Hopefully, future systematic studies will increase our knowledge of the effects of cat purring—on cats as well as on humans. Maybe purring is a meditative sound that cats can employ when they want to relax themselves or even pacify other cats.

Cats are born deaf and blind, but are able to perceive the vibrations of their mother's purring. It is how they find the milk they need to live. It is possible that cats communicate with their young by purring because it is a very soft sound that is not easy for predators to hear.

Young cats often purr when they encounter adult cats, so as to signal that they accept that they are at the bottom of the social hierarchy and that they have only peaceful intentions. The older cat often answers with a purr in order to make it clear that the young do not have anything to fear from them.

Many wild cats purr as well. Among the best known may be the beautiful cheetah Caine, whose sounds were recorded by Dr. Robert Eklund in South Africa. You can find an example on my website in the category "Purr(ing)," under the keywords "Caine the purring cheetah." Though there is a big difference in size between a cheetah and a house cat, they purr very similarly. The frequency usually lies between 18 and 25 Hz (so it is very deep) and the inhalation phase is roughly as long as the exhalation phase.

Research has determined that every kind of cat can either purr or roar, but never both. That lions, tigers, jaguars and leopards roar rather than purr probably has to do with the difference in the anatomy of their larynxes (voice boxes) to those of the purring cats. To be more precise, the degree of ossification in the hyoid bone beneath the tongue presents a plausible explanation for the fact that some cats purr and others roar. Roaring cats have an incompletely ossified hyoid bone, which enables them to roar but not to purr. All other cats have a completely ossified hyoid bone, which allows them to purr but not to roar.

One exception is the snow leopard, who appears to be able to purr despite having an incompletely ossified hyoid bone.

Some cats purr a little louder during inhalation, others during exhalation. Adult cats seem to purr the most when they are fed or petted. Some also purr when they are alone in their favorite place, like our Rocky, who often purrs for hours at a time when he is lying in his basket on the desk. Other cats start to purr when they arrive someplace new and want to investigate if it is safe. That is how it was with my black-and-white Fox, who walked around my apartment purring and getting to know everything the first time he was there.

If your cat does not purr, it is possible that they were separated from their mother too early and never learned to make the sound. It is also possible that they simply purr very quietly, so that we humans can hardly perceive it.

TIP: If you are having difficulties hearing your cat purring, try to make them lie down on your pillow next to your head and pet them until they start to purr. Even if the purring is very quiet, the pillow will act as a kind of resonance box and amplify the vibrations so that you will be able to hear—or feel—them better. The larynx, or even the whole body, vibrates when a cat purrs.

Anyone who is near a purring cat can hear that the sound is loudest at the front of the mouth, which indicates that it is produced in the larynx, or, to be more precise, with the vocalis muscle in the vocal folds (vocal cords), which produce the vibrations through a rapid twitching or contracting. Dr. Robert Eklund, with whom I conducted the research project "Melody in Human–Cat

Communication," has investigated the anatomy of the larynx of many cats, including wild cats, but even he can only speculate when asked how cats actually purr.

One possible means of investigating the articulation (position or movement of the mouth) of purring cats with phonetic methods would be to employ an ultrasound or MRI recording of the larynx, with which one could see which organs vibrate and how the vocal folds move. But how can you convince a cat to purr when it is in a strange room with strange people, tied up in a scary, loud device and then shoved into a tube? Or when a strange vet sticks a hard ultrasound sensor firmly against their shaved throat, which has been coated in gel? They will hardly be in the mood to purr!

Not all purring sounds are the same. The same cat can even produce very different purrs depending on emotional state, mood or situation. A study in England concluded that cats have developed a special loud kind of purring or a "cry embedded within the purr" that they employ when they want attention or something to eat from us humans (McComb, Taylor, Wilson & Charlton, 2009).

Cats can also combine their purring with other sounds. A cheerful and hungry cat can meow, trill and purr in sequence—usually in anticipation of a treat they are about to receive. Many cats purr and trill when they want to cuddle with their humans. A sleeping cat can combine purring with snoring.

CONCRETE EXAMPLES

Rocky the super purrer has a very calming purr, which he produces by slow inhalations and exhalations. Donna usually purrs a little more quickly, especially when she's happy about something, such as a treat or being able to crawl into my cardigan while I'm sitting at my desk. Turbo also often purrs when he is happy about something. He likes to combine meowing and trilling with purring, and sometimes his purring, which is usually so meditative, can even turn into a very agitated noise.

Kompis also likes to purr. His purr is often very deep and loud. He mostly purrs when we give him something to eat, pet him or when he is in his favorite place—the footstool with the soft blanket in the hall.

When I read about the English study with the title "The Cry Embedded within the Purr," I was not entirely convinced that such a purring cry or call really exists, as I had never observed it in my own cats. Can all cats really produce such a sound? I started to examine all the purring sounds that I had recorded of my own and of other cats with my phonetic ears, and although I have not been able to make out a call or a cry in the purrs of my cats, I did find that there are many varieties of purring and that cats can pep up their purrs with other sounds. Donna often trill-purrs or squeak-purrs, which are truly the cutest cat sounds that I have ever heard. She kneads around on my lap and purrs, trills and squeaks in ever quicker succession until the sounds are all mixed up—it is irresistibly sweet!

Turbo sometimes purrs in his sleep. As he also often snores, he sometimes ends up producing a combination of snoring and purring, which is also very cute! Have you also heard your cats make unusual combinations of purring and other sounds?

> **TIP:** At the end of the book you will find links to sound files of purring cats on the website. Maybe you will find some similarities to the purring of your own cats! See the appendix starting on page 237.

CORRESPONDING BODY LANGUAGE

Because purring is usually a sign of well-being, most people have a mental image of a cat curled up happily on the lap of its human. Indeed, the typical purring situation is when our furry friends are comfortable in their favorite place, whether it is the lap of their human, in a basket or maybe on a soft blanket. It is not uncommon for them to make a turning motion, often kneading with their paws in the process, before lying down. Kittens purr while they are at their mother's teat so as to stimulate the production of milk. Many cats also purr sitting, standing or walking. Eye contact with humans and ears pointed forward often go with purring. The tail is usually still, and often raised, and its tip can be curled like a question mark. This is a sign of affection and tenderness. It is even more intimate if a cat blinks or closes its eyes altogether while it purrs. If your cat blinks at you like that it probably means "I trust you from the bottom of my heart."

PHONETIC CATEGORIZATION (SOUND TYPE, MELODY)

Articulation

Purring is usually produced with a closed mouth, during inhalation as well as exhalation. The air is mainly inhaled and expelled through the nose, so purring can be classified as a nasal sound; it is, however, probably mostly a voiceless sound. Purring has no clear melody, unless it is mixed with trilling or meowing. It is also typical of purring that it is very quiet and often lasts for a long time (often several minutes). That is possible because the cat does not need to take a break between each purr in order to inhale. Instead, the sound can be produced while the cat inhales as well as while it exhales. The inhalation lasts somewhere between half a second and one second, and then quickly turns into the exhalation, which lasts about as long. In some cats, one of the phases can be significantly longer, louder and deeper in frequency.

Phonetic Description and Transcription

Purring is a soft, very deep, probably mostly voiceless (but regularly vibrating) consonant sound completely without vowels. It most closely resembles an airy, nasal trill like an [ʀ̃] or [r̃] often combined with a soft [h] consonant. Every inhalation and exhalation phase has a length (duration) of between one half and one second. The transition between phases is short and rapid.

Moelk transcribed purring as [ˌhrn-rhn-ˈhrn-rhn…]. With the help of the International Phonetic Alphabet, I would write this as roughly [↓hːr˜-↑r˜ːh-↓hːr˜-↑r˜ːh] or [↓hːʀ˜-↑ʀ˜ːh-↓hːʀ˜-↑ʀ˜ːh]. Purring can be combined or even mixed with other sounds, frequently trilling or meowing.

TIP: The phonetic symbols are described in detail in Tables 3, 4 and 5 at the end of the book (pages 260-265).

Voice and Melody

Purring is mostly voiceless, although it is probably produced using muscles in the vocal folds. It is a regular sound, but so low-pitched and deep that one can hear the pulsating of the larynx as it vibrates. It sounds almost like a softly rattling chain to the human ear. The very deep vibrations are more or less monotone, though there is a difference in the frequency of the inhalation and of the exhalation in some animals. Mixed purr combinations (with trilling and meowing) can be voiced.

8

"MEH MEH MEH MEH! I AM GOING TO EAT YOU UP!" CALLS FOR PREY

Rocky sits on the windowsill in the kitchen, looks out the window and makes a sound that one does not necessarily associate with a cat: *meh, meh, meh…eh, eh…eh, eh eow…* When I go to him, I see that there is a large gull sitting on the roof of the neighbor's house. Rocky's tail wags in large movements. Is he excited? Nervous? *Meh, meh, meh. Eh, meow*, he goes on, staring at the gull hypnotically. The gull does not pay him any attention at all, and eventually it just takes off into the sky. "Good Rocky, good boy," I praise him. "You've chased off another gull." Today, I know a lot more about the reasons for these strange cat sounds, but the first time that I heard them—I think it was from Vincent—I had to laugh.

DESCRIPTION OF THE SOUND

Chattering and chirping are fairly uncommon sounds that can be easily misunderstood by cat owners. Many of our cats like to sit on the windowsill and look outside. Then they discover a bird in a tree, on a roof or on the street, and start to chirp and to chatter. Some cats utter the same sounds outside in the garden, too, if they discover prey that is too far away, an insect or a bird, for example. Some wild cats may demonstrate the same behavior when their preferred prey (birds, rodents or insects) are nearby. These sounds could be part of the hunting instinct, as the cat is trying to imitate the sound made by the prey. Perhaps they want to lull the prey into a sense of security, so that they can stay hidden as long as possible.

As there are a great number of different variations, it is not necessarily straightforward to categorize the variety of these sounds. One could divide them into the following categories, bearing in mind that variations and even subcategories are possible: chattering (voiceless) and chirping (voiced). Which of the following sounds or variations of sounds have you heard your cat saying?

Chattering

Chattering is generally a number of voiceless, very rapid, short, clipped sounds that are uttered in stuttering or clicking sequences. They are produced with a rapidly juddering jaw. They often sound like teeth chattering with a *k* consonant, and can be written in the phonetic alphabet as [k k k k k k] or [k̟˭ k̟˭ k̟˭ k̟˭ k̟˭ k̟˭]. The small "+" under the *k* means that it is pronounced farther forward in the mouth (in the middle of the mouth), and the "=" means that it is an unaspirated *k*, that is to say, no more air is expelled after the *k*.

It has been suggested that this sound is connected to the capturing of prey. A cat who sees an unreachable bird chatters and imitates a killing bite in a stereotypical way. The action could serve as a means of stress relief. Some cats also chatter as a means of protest, for example when they feel they have been mistreated by their humans or when they are annoyed.

Chirping

Chirping consists of short, voiced sounds that sound almost like *meh* or *eh*, and resemble the chirping of birds or the peeping of rodents. Some also hear the ringing of a telephone in the sound. The tone (intonation, melody) often falls at the end of the sound. The beginning of the sound is often hard, with a glottal stop [ʔ], and the vowels are frequently *e* as in "men" and *a* as in "cat," or [ʔə] in phonetic notation. These sounds are often repeated in sequences like *meh, meh, meh.* Chirping can sometimes be a little hoarse, raspy or raw and sound similar to a hoarse meow or snarl. Besides being a hunting instinct, chirping can also occur in other situations, for example when a cat hears a lamp make a crackling sound after it has been turned off, or sees people throw small objects through the air, as my husband and I do when we throw darts at a dartboard.

In addition, there are variations of chirping that have somewhat different phonetic characteristics and might therefore be assigned to potential subcategories.

Soft tweeting is a mild form of chirping. There is no hard beginning (no glottal stop [ʔ]); instead it sometimes begins with a *u* or a [w] and the vowels vary, though *i*, *ae* and *u*, such as in [wi] or [ɦɛu], are typical. Tweedling or warbling is a prolonged chirp or tweet. It is often combined with voice modulations such as tremolo or quavering and has a much more varied melody. Two or more syllables are often audible, such as in [ʔəɛəɥə]. Here, too, combinations with the aforementioned sounds have been observed.

CONCRETE EXAMPLES

My cats are very different when it comes to this category of sound. Donna mostly produces combinations of chirping and chattering when she sees a bird. She seems to prefer large birds such as gulls, crows and magpies—the larger the bird, the louder and longer her sequences of chattering and chirping.

Turbo chatters only occasionally, but he chirps all the more—and not only at birds and insects. When my husband and I play darts in our little bar corner, there are small interesting objects flying through the air and Turbo is of the opinion that they might be something to hunt. And that is worth a chirp as far as he is concerned. To the eyes of cats, darts have a striking resemblance to the small birds that they know from the garden. Of course, we throw only darts that have soft tips and we take care that none of our cats cross in front of us while we are playing. Turbo does not seem to be ready to acknowledge that the flying objects are neither tasty treats nor birds and chirps at them every time our electric dartboard is in use. He runs to us very quickly and says [ʔɛ ʔə], [ʔæ ʔa ʔə], which we take to mean "What? You started without me?

You know that I want to watch. And where is my treat?" During the game he takes up residence on our bar and gives a chirping commentary to every throw.

Rocky chirps and chatters in combination, and he is the only one of my cats who also often tweets or tweedles. He can sit for a long time observing a bird while tweeting and tweedling softly to himself, with a rising and falling melody. He often does it when he sits on the windowsill and discovers a bird outside.

Maybe it is more of an individual variation, but I have introduced tweeting and tweedling as two further subcategories in this book so as to make the spectrum of these sounds somewhat wider and to help you describe the sounds your own cat makes with phonetic terminology.

Some cats chatter but do not chirp, others can chirp but not chatter and some do neither. For example, Vimsan almost never chatters. I do not know why. Maybe she caught so many birds before she came to us that it does not excite her so much when she sees a small bird through the window. She sometimes likes to climb trees on the hunt for magpies when they have teased and cursed her, but she does not make a peep in the process.

Corresponding Body Language

Cats usually sit or stand when they chirp or chatter. The sounds are often accompanied by a tail whipping back and forth impatiently. That shows how excited, tense and concentrated the animal is.

Sometimes cats try to catch their prey even though it is on the other side of the window. While chirping and chattering they snap their teeth, judder their jaws, and thus open and close their mouths in a quick repetitive manner. Some have claimed that cats do the same when they have caught a small animal and want to eat it (with fur and bones and such), juddering their jaws in the same way so as to protect their throats from sharp bones. Others surmise that cats are practicing the killing bite they will deal to their prey when they chatter their teeth and judder their jaws. Cats can also chatter when they are lying down and when they sleep. Turbo probably sometimes dreams of birds or darts and chirps and chatters softly when he is sleeping in his basket.

PHONETIC CATEGORIZATION (SOUND TYPE, MELODY)

I have observed tweeting and tweedling only in my own cats, as I have already said, and have rarely heard it in others. There are certainly other variations, perhaps some of them are present only in your cats, so of course I cannot account for them in my description. The diversity and the possibility for variation in cat sounds are almost endless. This is one of the reasons for my fascination with the sounds of cats.

Articulation

Chattering and most variations on chirping are produced with a mouth held tensely open. In contrast, the softer subcategories of chirping—tweeting and tweedling—are usually expressed with an opening, closing or opening-closing mouth.

Phonetic Description and Transcription

Chattering sounds are often produced using a series of identical consonants and sound almost like glottal stops [ʔ ʔ ʔ ʔ] or like clattering *k*'s produced at the front of the mouth [k k k k k] or like [k̟˭ k̟˭ k̟˭ k̟˭ k̟˭ k̟˭]. The small plus under the *k* means that it is pronounced farther forward in the mouth (in the middle of the mouth), and the "=" means that it is an unaspirated sound, so that no air is expelled after the *k*.

Chirping often consists of [ʔ] or [k̟˭] and a vowel such as *e*, *ae* or *a*, such as [ʔə] or [k̟˭e]. Chirping usually occurs in longer sequences, such as [ʔɛʔɛʔɛ].

Tweeting and tweedling both occur without an initial consonant and begin with a soft *u* [w] or *h* instead, such as [wi] or [ɦɛu]. Longer tweedling is often pronounced without [ʔ] or [k̟⁼] and consists of multiple syllables that often produce a fairly complicated melody: [wəɛəɥə].

TIP: The phonetic symbols are described in Tables 3, 4 and 5, starting on page 260.

Voice and Melody

Chattering is mostly voiceless, whereas chirping is voiced. The short chirping sounds are usually either monotone or have a melody that declines slightly. Tweeting and tweedling in contrast can have more variation in their melodies. Tweedling consists of a combination of sounds with numerous rises and falls in the melody.

9

LEARNING TO UNDERSTAND YOUR OWN CAT

Now that I have presented numerous cat sounds and their variations, I would like to use this chapter to categorize these according to a larger phonetic system. In the process, something like a verbal scaffolding for the language of cats will be produced. The different kinds of sounds will be presented in a table, so that the phonetic characteristics can be easily recognized. Furthermore, I will name more variations and will attempt to give possible reasons for these variations.

SUMMARY OF CAT SOUNDS: THE SYSTEM OF CAT LANGUAGE

If we forget the articulatory-phonetic categories that Moelk introduced (see page 31) for a moment and try to approach cat sounds using auditory (i.e. careful listening) and acoustic (analysis of acoustic patterns related to frequency, length [duration] and intensity [loudness or volume]) evaluations, we can categorize the various sounds that cats can produce according to a phonetic system and try to compare them to the sound systems of human languages such as English. As I have mentioned before, it is important to remember that the sounds made by cats cannot be directly compared with those existing in human speech.

Although cats communicate with their humans in a complex fashion, I have not yet found any indication that cat sounds follow grammatical principles or that every sound or type of sound can be translated one-to-one in a specific word or sentence in human language. That is not to say that cats cannot express their feelings, moods, wishes and needs with sounds. They most certainly can. But every cat develops a system—together with their humans, as well as possibly with cats they are close to—that allows them to communicate in a unique and special way. Every cat uses multiple forms of communication (with scent, with tactile signals, with visual signals, such as body postures or ear movements, and with sounds). In the end, they seem to choose the form of communication that best serves their needs and continue to apply it in similar situations.

Sounds often seem to be the preferred means of communication with humans, and meowing has proven to be especially effective as we react to it immediately. But *meow* is not a word, as it does not have a unique and unambiguous meaning. The cat communicates its needs and desires to us using different voice qualities, melodies, volumes (acoustic sound level pressure or intensity), and combinations of vowels and consonants. We humans learn, too, with a little practice, to understand these nuances and can even use them ourselves when we speak to our cats in human language.

A quiet, soft, and bright (with acoustically high resonances) voice usually indicates friendliness and affection, whereas a loud, hard, and deep voice shows that we are unsatisfied or angry—regardless of which words we use. Intonation, the rise and fall of the pitch (melody), for example, often expresses more than the words themselves. We humans also often communicate wordlessly with our close friends and family members. An expressive *mmmmm*, for example—pronounced with various melodies, lengths or volumes—can have a number of different meanings.

Although cats do not communicate in a language that is similar to human languages, I would still like to try and organize the sounds that cats are able to produce using a phonetic system. This would contain all the vowels and consonants that I have observed so far, as well as additional phonetic characteristics. I'm interested to know which vowels and consonants cats can actually pronounce and in which combinations they occur.

The system is far from perfected—I have not yet investigated, for example, whether cats use nasal vowels, such as in the French *un bon vin blanc* (a good white wine). Still, I would like to present my results so far.

For an overview of the cat sounds that I have described in this book, you will find Table 2 at the end of the book, starting on page 253, that summarizes all of the sound types and their corresponding phonetic characteristics. The table presents all of the various designations for the sounds as well as their subcategories, the type of articulation (position or movement of the mouth), the register of the voice (high or low pitch), the typical phonetic transcription, as well as some additional comments—all at a glance.

Not all nuances and variations are contained in this overview, which presents only the most important categories and types of sounds. Nevertheless, I believe that you can use the table as a starting point when you hear a cat sound and are not sure what it could mean. The sounds are categorized by type—meowing, trill-meowing, trilling, growling, hissing, howling, growl-howling, snarling, mating calls and chattering, as well as by the subcategories of those types. Under Articulation, you can find out whether the sound is made with a closed, open, opening and/or closing mouth, while Voice indicates whether the sound is voiced or voiceless, whether the pitch is high or low, whether the melody rises or falls. A brief description of the sound can be found under Phonetic Category, and Typical Phonetic Transcription represents the sound in phonetic characters. There are also some additional notes on the sounds.

Just like in human speech, most cat sounds consist of more than one individual sound (vowel, consonant). The English word *meow* consists of one consonant sound, the *m*, two vowel sounds, *e* [i], *o* [a], and a semivowel sound, the *w* (the *w* is called a semivowel because it sounds very similar to the vowel *oo* [u] in the Indian or Irish English pronunciation of *foot*). The cat sound [miaw] also consists of one consonant, two vowels and a semivowel. These building blocks provide hints about the anatomy of a cat's vocal tract and the movement that a cat can make with its mouth, tongue, lips and vocal folds (vocal cords).

Cats have much higher voices than humans. That is because they have much smaller vocal apparatuses. Smaller vocal folds produce higher pitches (and tones of voices) and smaller resonance chambers in the mouth produce higher (brighter) sounds.

It should not be assumed that cats can produce every individual sound that is present in human speech. The lips and tongues of cats are differently shaped and of a different size than those of humans, and the larynx (voice box) is also differently shaped and placed. Therefore we cannot imitate every cat sound precisely either. Can you purr, trill or growl without a great effort, for example?

Let us begin with the smallest building blocks of speech, namely individual sounds—vowels and consonants.

Vowels

So far, I have discovered more than ten vowels in cat sounds. In the graphic that follows, I have transcribed these vowels using phonetic symbols and arranged them in a kind of phonetic space. The vowels are arranged according to tongue position (height, frontness) and lip position (unrounded/spread or rounded) in a vowel diagram or vowel chart, also sometimes referred to as a vowel quadrilateral.

The figure on page 150 shows the vowels of human speech that are included in the International Phonetic Alphabet, the vowels I have observed cats produce so far (circled), as well as those that I think cats can produce (in dotted circles). When two vowels are arranged on the right and left sides of a dot, the vowel on the left is formed without lip rounding, and the right is formed with lip rounding. Otherwise the pronunciation is identical.

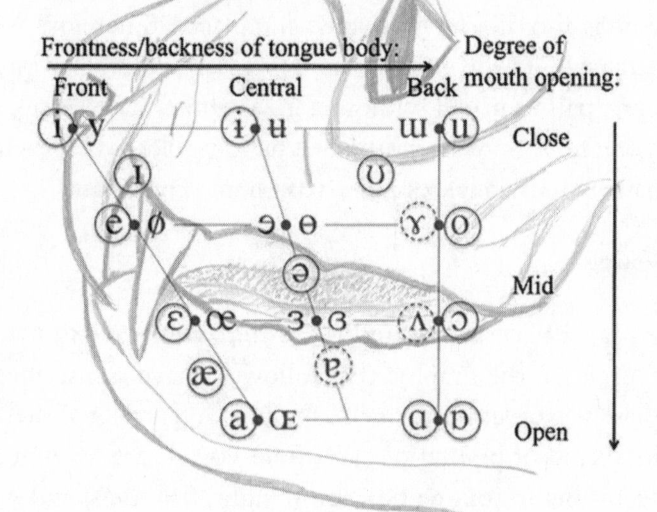

Vowel chart with phonetic characters for all vowels (vowels observed in cat sounds have circled borders).

Although I have been observing cats for a long time, I still cannot say with certainty whether they can protrude and round their lips, as we humans do when we go from saying *u* [ʌ] in *strut* (spread lips) to saying *o* [ɔ] in *thought* (rounded lips). They should be able to do so. All mammal infants do it instinctively when they nurse, and the *w* in meow is a rounded semivowel according to the phonetic descriptions and can be produced with rounded protruding lips. Until now, however, I have not recorded any cat who clearly and visibly protrudes and rounds their lips. Cats can, however, produce vowels with both a more open (for instance when producing the sound [*a*]) or a more closed mouth (when producing the sounds [*i*] or [*u*]). I assume that most of the vowel sounds I have observed in cats are produced without rounding the lips. It also seems likely that cats may produce sounds with neutral—neither rounded nor spread—lips, such as [ə] (like the *a* in *comma*) or [ɐ] (like the *o* in *lot*).

Consonants

My research is based on the assumption that cats can produce consonant sounds similar to the sounds made by humans. I have summarized all of these consonants using a phonetic system. Occasionally, I have had to make compromises, such as with the trill, similar to an *r*, that occurs in purring and trilling. It is unlikely that this is a rolled *r* as it would be articulated by humans. Instead, it is probably produced farther back in the mouth. Even so, I have transcribed this sound as [r̃] in phonetic script, because it is brighter than the growling sounds transcribed with [ʀ̃].

There are much larger differences between different consonants than between different vowels, and therefore we phoneticians tend to subdivide them into different consonant categories. I have observed the following phonetic consonant categories in cat sounds.

Stops are produced by first completely blocking the vocal tract so that no air can escape from the lungs, causing a slight pressure difference to be built up behind the closure, and then by quickly opening the closure, causing the released airflow to make an audible sound (a short burst). I have identified the stops [t], [c], [k], [ɡ] and [ʔ] in cat sounds like chatters, chirps and spits.

Fricatives are characterised by a turbulent airflow causing frication noise. They are produced by forcing air through a narrow constriction made by two articulators close together, e.g. the upper teeth against the lower lip as in [f]. [f], [ʃ], [ʂ], [ç], [ɦ] and [h] are among the fricative sounds that I have observed in cat sounds like hisses and spits.

Approximants are consonants produced by two articulators approaching each other closer than with vowels, but without causing any turbulent airflow. Some approximants are so similar to vowels that they are sometimes called semivowels. I have observed the approximants [w], [j], and [ɥ] in meow-like sounds, and [ɹ] in trills.

Laterals are *l*-like consonants which are produced with

the tongue against the middle of the palate, allowing the airflow to proceed along the sides. Cats should be able to produce laterals, but I have heard only a few instances of *l*-like sounds in cat vocalizations, and have not yet been able to record any such sounds myself.

Nasals are consonants produced by a closure in the oral cavity or at the lips, but with a lowered soft palate (velum), allowing the airflow to escape through the nose. In cat sounds, trills and meows sometimes begin with an [m], and in the transition from a trill to a meow, sometimes a [ŋ] can be observed.

Trills are produced with an airstream causing an articulator to vibrate, causing a regularly pulsating sound. Cat trills (chirrups, grunts, murmurs) often contain [r] or [ʀ]-like sounds, which are usually nasalized [r̃] or [ʀ̃], as the mouth is closed.

Affricates are combinations of one stop consonant followed by a fricative produced with the same articulator. In cat sounds, spits are sometimes characterized by an initial affricate, such as [t͡ʂ] or [k͡h].

Table 1 on page 155 shows the consonants which I have observed in cat sounds, based on the assumption that cats are able to produce consonant sounds which resemble human consonants. The table rows contain the consonant categories described above, and each column corresponds to the specific place of articulation of the consonants:

bilabial—with upper and lower lips (as in [m], [p] and [b])

labiodental—with the upper front teeth against the lower lip

dental—with the front of the tongue against the teeth

alveolar—just behind the teeth

postalveolar—somewhat farther back than alveolar

retroflex—farther back with the tongue between the alveolar ridge and the hard palate

palatal—with the tongue against the hard palate

velar—farther back with the tongue against the soft palate

uvular—even farther back with the back of the tongue against the uvula

pharyngeal—with the root of the tongue against the back of the throat (pharynx)

glottal—all the way back in the vocal folds

	bilabial	labiodental	dental	alveolar
Plosive				t
Nasal	m			
Trill				r
Fricative		f		
Approximant				ɹ
Lateral				l(?)
	postalveolar	**retroflex**	**palatal**	**velar**
Plosive			c	k, g
Nasal				ŋ
Trill				
Fricative	ʃ	ʂ	ç	
Approximant			j	
Lateral				
	uvular	**pharyngeal**	**glottal**	
Plosive			ʔ	
Nasal				
Trill	ʀ			
Fricative			h	
Approximant				
Lateral				

Table 1. Consonant table with phonetic characters for consonants observed in cat sounds. The "(?)" indicates that I assume that cats are able to produce a sound which I have not yet recorded.

In addition to the consonants depicted in Table 1, cats are probably able to produce the following consonants:

[w] (voiced labial-velar approximant)

[ɧ͍] (voiceless simultaneous postalveolar and velar fricative, produced with an open mouth and spread lips)

[ɥ] (voiced velar approximant)

The table of consonants at the end of the book has all of the consonants that I have discovered so far in my recordings of cat sounds.

Additional Phonetic Characteristics: Prosody

Cat sounds do not simply consist of building blocks like vowels and consonants. They also have an intonation, an intensity, a length, a rhythm and a voice quality (for instance a harsh, breathy or creaky voice). When you put these characteristics together, you get the prosody. Just like humans, cats are capable of distinguishing between a meow that is supposed to mean "I am hungry" and a meow that should mean "I do not like that." That is why I believe that we humans can interpret cat sounds much better when we concentrate more on their specific characteristics. Is a sound long or short? With the [ː] symbol we can indicate length. Without the symbol the sound is short, but with the symbol the sound is long. Sounds followed by the length symbol are long, such as the *a* in [wa:uh]. Changes to the prosody during a cat sound—such as rising pitch or increasing volume—also transmit important information. We still do not know much about the variation in the vowels, consonants and the prosody in cat sounds, but it seems likely that cats are able to vary their voices and their intonation just as other animals including humans do, and we know a bit more about this general variation.

In order to understand cat sounds even better, we must begin by investigating the characteristics that are common to all mammals—whether cat, human, hippopotamus, gorilla, wolf or mouse. According to the "frequency code"—the cross-language and cross-species use of frequency—first described by the American professor of linguistics John Ohala (1994), low-pitched and dark (with acoustically low resonances) sounds are used to signal "I am big, strong and dominant." Such sounds (sounds with low frequencies) are often used in aggressive situations. High-pitched and bright sounds often signal the opposite—"I am small, weak, friendly and submissive." Young animals (and people) have more high-pitched and brighter voices than adults, and dogs yelp or whimper when they are uncertain or submissive, sad or afraid, but often growl or bark when they are threatening or aggressive.

In addition, some vowels often sound big, dominant and declarative (*a*, *o*), while others often seem small, weak and questioning (*i*, *e*). These vowels occur relatively frequently (though not exclusively) in words with corresponding meanings: *large*, *tomcat*, *strong* (big, dominant, aggressive); *teeny*, *kitten*, *weak* (small, friendly, peaceful). This occurs in many languages, including Swedish and German: *sort*, *groß*, *stark* (large, strong); *liten*, *mini*, *niedlich* (small, cute). We should also note that the intonation or melody can be influenced by emotion, or by a specific intention. Uncertain speakers (who need confirmation from their conversation partners), often end their utterances with a rising, questioning tone (intonation, melody), whereas self-confident speakers tend to employ a declarative, falling tone.

Many cat owners claim that they can immediately hear whether their cat is content (happy), sad, angry or afraid. It is possible that animals send out the same signals as we humans in order to express their emotions and needs. However, as cats are predators, they do not often express their feelings. It is a way of protecting themselves; they do not want to reveal any weakness or that they are in pain.

As a rule, we humans reveal our feelings with our voices. Regardless of the language we speak, we change the melody, volume and tone of our voices when we are sad, happy, afraid or angry.

Joy: When we are happy, our voices are high-pitched, and we employ a full-bodied sound with a wide range of frequencies. Moreover, the intonation is characterized by numerous large and rapid changes in pitch.

Anger: When we are angry or enraged, we often speak in a fairly high-pitched voice, very loudly and with abrupt changes to our voices (due to the increased muscle activity in our speech organs). However, we may also repress our feelings and speak with a quiet, somewhat slower, pressed or tense voice. So there are at least two ways in which we can express anger.

Sadness: Soft, slow and low tone of voice—these are the signs of a sad voice. The range of frequencies remains small (only small variations and a minimal difference between the highest and the lowest pitch), and so the intonation is rather monotone (due to minimal muscular activity in the speech organs).

Fear: When we are afraid, we take long pauses while we are speaking, although we speak in a rapid tempo with a wide range in the melody. Irregularities may also occur in both the frequency and the intensity.

With the help of these universal features of prosody, we can better interpret the sounds our cats employ. Cat sounds, too, sound different depending on the cat's mood or emotional state. I have noticed that my cats often change the melody at the end of the sound. The question arises, what happens to make one meow rise at the end, while another falls? That is why I have observed my cats systematically, and have made both audio and video recordings of the situations in which they use different melodies. These confirm my assumption that cats, exactly like we humans, can express their feelings with their voices.

When Donna wants to persuade me to come and play with her, and does not immediately get the reaction she wants, she changes her voice. Her trill-meows get increasingly louder, slower (more extended) and are accompanied by larger changes in the melody. So every meow ends in a higher pitch than the meow before.

Kompis does exactly the same when he wants to be let out and we do not open the door immediately. His bright and high-pitched meows get progressively more high-pitched and bright towards the end. When Turbo is packed into his carrier when he needs to go to the vet, his meows sound entirely different from when he demands a treat at home. In the carrier, his meow sounds are very anxious and almost sad. They have a narrow melodic range, the voice is subdued, and the tone often declines towards the end.

VOCAL EXERCISES FOR HUMANS

As you have learned (and already knew from experience), cats can make a great variety of different sounds. Purring is different than meowing, hissing is separate from trilling and chattering is something other than howling. But two trills can also be different from one another. When I listen to the sounds made by my cats, I discover a great number of different nuances. Donna uses more *ae* [wʊæ] in her meows when she trills while kneading around on my lap than when she stands at the door, trill-meowing because she wants to be let out. Then it sounds more like a "typical" meow [wɪaʊ]. With a little practice, anyone can learn to recognize these nuances and variations and will understand their feline friend better. A little trick that I often use myself is imitation. Whenever I hear an unusual sound, I try to imitate it. Then I can feel how I move my mouth, lips and tongue in order to produce a similar sound. Maybe cats make similar movements with their vocal organs.

The accompanying body posture and movements of both the individual body parts and of the entire body can also reveal much about the probable meaning of a sound. Look very closely at what your cat is doing when it makes a specific sound, and you will be sure to see that verbal and visual communication reinforce one another and the message being conveyed will be clearer.

10

HUMAN TO CAT—
HOW COMMUNICATION SUCCEEDS

By now, you have read a lot about the various cat sounds, how they can be described and which phonetic features they demonstrate. Many cat fanciers who hear about my research into cat sounds ask me whether I have already cracked the cat code and can understand *everything* they say. Of course I haven't, but with a little practice, any cat lover can learn to understand the vocal (acoustic) signals used by cats better, and then improve their human-to-cat communication. In the following section, I have compiled some examples that reveal where I met with success, combined with some practical tips to help you better understand your cat.

HOW TO BETTER UNDERSTAND THE SOUNDS MADE BY YOUR CAT

Sometimes, the simplest of phonetic methods can help you to better understand a sound made by your cat. Try to listen precisely to how a sound is formed. Maybe you will discover the individual vowels or consonants contained in the sound. Try to imitate the sound yourself so that you can better understand how it was produced—with a closed, open or opening-closing mouth. Listen very carefully for changes in the melody, how short or long the sound is and so on.

Try to describe the sound, either with phonetic symbols (you can use the tables that you will find at the end of the book) or with a brief note about what you have heard, for example, "a long and strange sound that resembled a meow, though the mouth seemed to be closed most of the time. Proper vowel sounds were only observed at the end of the sound (*a* and *u*)" or "repeated very short *k*-like sound." You need not be a phonetician to describe a sound well. The main thing is that you describe the sounds clearly, so that you can understand them later. You should also describe the situation in which the sound was uttered—whether it was morning or night, in the kitchen or in the garden, whether you were playing with your cat, or your cat had just entered the room. Try to find out whether your cat uses similar sounds in similar situations. It will be terrific if it works, as you will be able to understand a sound much better if you can associate it with a particular situation.

But there are certainly some situations in which you simply do not know and cannot figure out why your animal reacts in a certain way or behaves so strangely. I would like to use this chapter to address some problem areas that cat fanciers have presented to me.

Why Does My Cat Not Say Anything?

Sometimes cat owners ask me whether I know why their cats scarcely meow or make other sounds. In return, I always ask them whether or how often they speak with their cats. Frequently, the answer is either "rarely," or "well, she seldom meows, but whenever she does, I always quickly say, 'be quiet.'" I have confirmed it countless times, if we speak to our cats frequently, then they "speak" to us a lot as well. If you want to have a quiet cat, who prefers to communicate with visual signals, try not to speak with her too much. Use visual signals and touch instead.

Why Does My Cat Make Such Funny Noises?

When people ask me to interpret the sounds made by their cats, they sometimes say in advance that the sounds are very strange or unusual. Some cats can imitate sounds made by other animals, or even by their people, up to a point. It can be something like a lamp that makes a little clicking sound after it is turned off, which the cat interprets as the sound of an animal of prey and therefore answers with chattering. As already mentioned, our Turbo chirps at the darts when they fly through the air. Some cats also try to imitate the voices of their humans.

A little while ago, I received a question from a woman in the United States who wanted to know why her cat meowed with such a wonderfully and unusually deep voice that it almost sounded like a dog barking. I asked her to make a video recording of her cat, which she proceeded to do. She sent it to me a few days later. On the film, which her husband had shot, the woman could be seen as she was feeding her cat. She spoke to her cat as she did so: "Right, here we go, here we go," "Yes, here's the can, it's just a can," and "Yes, you're the one who wanted it, yes, dear." My phonetician's ears immediately perceived that the woman spoke with an unusually hoarse and rough voice. I wrote back that it was possible that her cat was imitating her voice, and that is why it barked so roughly and hoarsely. She wrote back right away and said thank you. She had not thought of it herself, but found it fascinating that cats could imitate the voices of their humans so effectively. If your cat makes sounds using a strange or unusual voice, it is possible that they are imitating either your voice or the voice of another person around them or in their family.

Why Does My Cat Answer When I Speak to Her, and What Is It Supposed to Mean?

Cats make specific sounds in specific situations. Sometimes they meow at us without any recognizable reason though. Meowing is a vocal signal that often catches our attention. Although we do not always understand exactly what they want from us, we often catch on faster if we conduct a dialogue with them. When you are talking to your cat and get an answer that you do not really understand, try this simple method: answer with a similar sound. Try to imitate the cat sound using the same melody, and see what happens.

If the cat meows again, then answer and observe the visual signals (posture and movement) they produce. Your cat might show more exactly what they want. Maybe they will look at the door to the garden, will run to their empty food bowl or will simply sit on the floor, looking at us with wide eyes. If we repeat such dialogues with our cats often enough and pay attention both to the visual signals and to the nuances in the meow-sounds, we can learn more about the various sounds that are preferred in different situations, and will be better able to interpret the sounds our cats make in the future. However...it is possible that cats are just bored sometimes and just want to have a conversation with us.

HOW TO COMMUNICATE BETTER WITH YOUR CAT

Do not misunderstand me. I do not think that we should communicate with our cats solely using cat sounds. As a rule, they also understand our human speech very well. But from time to time a situation arises in which it is better, faster or easier to communicate using cat sounds. I would like to demonstrate with a few examples.

Cat Diplomacy

When we see that our cat is involved in a physical confrontation with another cat, our first impulse may be to try to rescue them and avoid the worst.

I would not necessarily recommend that we humans get involved and try to bodily separate the contestants. It is very likely that we will just end up injured as well, bitten or scratched on the hands or arms. Then we will be in a much worse position to help our cats. Cat behavior guides sometimes recommend trying to scare the animals so that they will stop fighting and run away. Common suggestions include clapping your hands, loudly yelling no or throwing a pillow near (but not directly at) the squabblers. I have tried every variant of these suggestions. Sometimes it works, sometimes it doesn't. The most successful method I have tried is the acoustic approach: namely, by hissing. I stand one or two meters away from the cats and hiss loudly at them—that is to say, I imitate the hissing of a cat. Sometimes it works right away, and sometimes I have to hiss two or three times, but up until now it has always worked. The cats react to the hissing and startle, they separate, and either one or both cats run away quickly—often it is the opponent who retreats, as my cats recognize me and stick around.

Greeting Cats

Do you also have a special "greeting ritual" that you have developed and that you use with your cats in the morning or when you first see your cats after getting home at night? A way of maintaining your relationship or of showing each other how much you have missed one another? When you encounter a strange cat, or one you just do not know very well, they can be better greeted by imitating the typical movements that cats use in greeting. Get down on your haunches or sit down, so as to make yourself smaller. Do not turn toward the cat, but sit at its side instead. Do not look at it directly. Try to imitate a friendly cat tail with your arm and hand by bending your elbow, raising you lower arm and shaping you hand like an arch or question mark. With a soft, small voice, you can then try to speak to them. The frequency code comes into play again here: high-pitched and bright (with acoustically high resonances) voices and sounds count as friendly, dark (with acoustically low resonances) and deep voices as aggressive. Sometimes I try to imitate a soft, bright trill, a chirrup, with a rising melody, "Brrrrrriuh." Many of the cats I greet this way approach me, so that I can very slowly extend a hand and allow them to take a sniff. I might even have a treat with me that the cat can then have.

No, That Is Not Allowed!

Sometimes cats do things that are dangerous or that we, for whatever reason, do not like. A soft "No, darling, I already told you that is not allowed" will not do much. It is more effective to growl long and deep, *grrr*, hiss sharply, *hsssshhh!*, or spit, *kshhht!!*, with the corresponding body language (make yourself large). That works much better for me. I even have a sound for "no" (a hissing sound), and another for "go away" or "come in!" I then follow behind my cats, almost like a sheep dog, and make a special clacking sound (by clicking the tongue). My cats figured out very quickly that they should go away, or come inside, when they hear this sound. I would like to emphasize that I have tried this only around my own cats in my own space. There might be problems if I were to hiss with other cats. Hissing should be carefully considered. It would be best to speak to your vet, cat psychologist or therapist before employing aggressive cat sounds!

Calming Cats

Although many humans find it difficult to imitate purring, I have determined that my cats are less stressed when I try to purr—they lie down, stay down and close their eyes. I sit next to them, pet them slowly, and practice my purr as softly and as slowly as I can. I might have gotten a little bit better, but I still find it difficult to imitate this sound. But when you speak in your own voice with soft, low tones, it can be just as effective.

11

CAT PROBLEMS IN DAILY LIFE

As you have certainly noticed already, it is not just peace and happiness in our home. We have also had problems and misunderstandings with our cats. Some simply resolved themselves. Others we had to work on and actively look for solutions.

When our first cat, the beautiful, lovable Fox, moved in with us, I knew next to nothing about cats. At the beginning, I made a lot of mistakes, which I still do on occasion, by the way, which is why we still have problems sometimes. Over the years I have also learned a lot—from my own experience with cats, from books and from other media. I learn something new about having a cat, or about communication either among cats or between cats and humans, almost every day. I believe that every relationship has its problems from time to time. The challenges with the cats that we have confronted together have strengthened our relationship.

For example, when Vimsan had been with us for two years and had settled in well, she started going out more often and staying out longer. The triplets were not at all happy that she stayed away so long and brought strange smells from the neighborhood back with her. I have to admit that we did not even notice at first. Both my husband, Lars, and I were much too occupied with our work and were very tired in the evening. But when Rocky and Turbo started to follow and attack Vimsan, and I started having to clean urine stains off the floor, pillows, rugs and blankets—even off the wall behind the stove—every day, we were forced to admit that our happy family was not so happy anymore.

In this chapter, I will share my experiences with such problematic interactions with cats. Maybe they will be helpful to you in similar situations. I must emphasize that I am not a veterinarian, a cat psychologist or a therapist. There are no generally applicable rules with cats. Instead, it always depends on the context, the cat(s) and the specific case. I myself have often consulted with professionals. With their support and my own experience, I have been able to resolve many problems with my cats.

MY CAT ONLY COMES TO OTHER FAMILY MEMBERS, NEVER TO ME

When Vincent moved in with us, he showed a great preference for me. He lay purring in my lap for hours every evening. In the morning and the evening, he often followed me meowing so as to make it clear to me that he was hungry. My husband, Lars, often watched jealously; although Vincent let himself be petted by Lars, he never came and lay in Lars's lap. We often talked about this and concluded that it had to do with me having the most interaction with Vincent in situations where it was about his needs. I always fed him and brushed him, cleaned his kitty litter and smoothed his blankets when they were crumpled, as well as doing other things for him. I also played and spoke to him the most. So maybe it was understandable that Vincent tended to come to me when he wanted something.

Then I took a long trip with my colleagues (we spent two weeks at linguistics conferences and symposia in Malaysia and Singapore). Lars was home alone with Vincent at the time.

It turned out that Vincent understood that Lars was a really nice guy as soon as I was away for a few days. Lars fed him just as well, and he could relax on Lars's lap just as well. When I got home again, we set it up so that I fed Vincent, brushed him and cleaned his kitty litter in the mornings (because I am often up before Lars anyway) and Lars does the same in the evening. After that, Vincent became a very just, unbiased cat. Every night he spent an hour on my lap, and then exactly as long with Lars. At night, he often lay between us in bed, and turned to Lars just as much as to me when he wanted something or had something to show us. Finally, we were convinced that Vincent cared for us equally.

TIP: If your cat ever goes to only one person in the family, try dividing up the chores (feeding, playing, cleaning the kitty litter) among family members, so that the cat understands that they are loved by and can expect help from everyone in the family. And if your cat goes only to other family members and not to you, try to spend a few days alone with the cat and pay special attention to the cat in this time, play a lot and spend a lot of time with them. If you are lucky, you and your cat will have a much better relationship as a result.

MY CAT WAKES ME UP EVERY NIGHT

For a long time after he came to us, Vincent had a bad habit. In his old home, he had lived with a very intelligent roommate, a female cat who harassed him daily until he developed a bladder infection because he was so afraid to go to the litter box and hid under the sofa all the time. To help Vincent, we decided to bring this tyrant (a beautiful gray tabby named Kisseson) home with us for a while, until she could be returned to her former owner where she had lived before she got Vincent as a roommate. Then, after a while, Vincent came home with us as well, but by that time, Kisseson did not live with us anymore; we loved her a lot and it was very hard on me when she returned to her old mistress after a few months.

In the meanwhile, alone at his old home, Vincent had gotten tremendously fat, and his vet had put him on a strict diet. That means that he only got a specified amount of special food at a certain time. As a result, he was often hungry. He started to meow very early in the morning and jumped on our bed, trilling and purring, kneading the blankets, and doing everything he could to wake us, so that he could have his breakfast. He walked across our pillows and faces, sat on our chests, and looked at us intently, and if we still did not wake up, he bumped his head into our noses until we really could not sleep anymore. It was not a nice way of being woken up, and we felt no desire to be woken that way morning after morning. What to do?

I learned from multiple books and television programs about having pets how important it is to be consistent and not to give in if the animal wants something that is not good for it or that the human does not like. I said to myself that we had to practice consequences, so we employed the following method: we delayed dinnertime until shortly before bedtime, and we also played with Vincent for an hour before we went to sleep. He loved cardboard boxes and paper bags, for example, and used them as play houses before biting them to shreds. He also liked to play with his favorite toys, fur balls (preferably black), which we threw to him and he kicked through the apartment like little footballs. Then Lars and I went to sleep and promised each other to ignore Vincent when he woke us up at four thirty the next morning.

The next morning our alarm clock woke us, and Vincent came to us for his breakfast only when we got out of bed. Finally, we could sleep peacefully at night again. Of course we continued with our new nightly ritual of a late dinner and playtime. Vincent tried to wake us up early only a few times after that, but when we did not respond (instead we consistently acted as if we did not hear or notice him), he lay back down pretty quickly and waited until we got up.

We learned from this experience and we used this method from the beginning with the triplets, and never fed them immediately after getting up, but waited for an hour or two instead. And really, they have never woken us in the early morning so far. Donna just trills softly in the early morning in winter on occasion, because she wants to get under the blankets where it is warmer.

TIP: If your cat meows and wakes you up every night or very early in the morning, you can try changing its feeding time, too, and spend half an hour before bed together—play or cuddle for a bit. You, too, should stay consistent. Do not get up at three thirty to feed them (unless the cat is old or sick). I hope you will be as successful with this method as we were.

MY CAT IS TOO FAT

Turbo is the youngest of our triplets. The woman from the local humane society found him a day after his siblings, all alone and mewing weakly but desperately in a hedge near the community garden where the mother cat had brought her young. His mother had only a little milk for him.

His siblings, Donna and Rocky, were much larger and often pushed him away when he wanted to nurse. So he received extra nourishment (first formula, and then real cat food). When the triplets came to live with us after three months, little Turbo was the one who was hungriest, and that is how it stayed. He not only ate up his own food, he also went to his siblings' bowls to clean them out afterward. At first we did not notice that he was slowly gaining weight, as he was still the smallest. But when the triplets were grown we could no longer ignore the difference between Donna and Rocky, who were both slim, and the rotund Turbo. Even our friends had started to call him "Fatso."

We bought him diet food and took away the other bowls as soon as Donna and Rocky had eaten. But Turbo continued to get fatter. So we went to the vet with him and consulted a dietary specialist, who suggested that we put him in his carrier and leave him in the garden for an hour, so that he could gather some new experiences and be a little less fixated on food. At the time, we did not let the triplets out yet, as we had friends who lived on the other side of the street who had lost two free-range cats to car accidents. There is no way we wanted to experience that. So we played more at home with him, and tried to stem his hunger by putting some of his food in "food toys," like a cat activity "snack ball" and a perforated empty toilet paper roll whose ends we had sealed with paper. But Turbo still did not lose weight. He even started to steal his siblings' food before they were done eating. It couldn't go on that way.

Maybe we could take little walks with Turbo on a leash? I bought three cat leashes, and slowly got all three used to wearing their harnesses (of course we had to make Turbo's the biggest!) and then we started attaching the leashes. After that I started to take short walks in the garden with one cat after the other. Not with only Turbo, as I did not want to be unfair to the other two; rather, I wanted to give all three the chance to catch some fresh air and experience our garden. At first it was a lot of fun. Donna was always the brave one, who liked to walk and run fast and to crawl into our hedge, so that it was sometimes hard to get her out as the leash got stuck on the twigs.

Rocky was very cautious, but he seemed to enjoy slowly promenading through the garden, smelling the flowers and trying to catch butterflies. Turbo seemed to like being outside, too. He quickly learned how to walk around our house and found his favorite places (the bench in the sun, the flower bed in front of the greenhouse). The disadvantage was that it took over an hour when I took all three cats for a walk, and I often simply did not have time.

The cats also always wanted to stay out longer and longer, and meowed and pushed so as to be the first to get to go out, because they liked it so much. I did not know where I was supposed to find the time. Walking the cats on a leash for hours every day was not an optimal solution.

My husband and I thought about solutions for a long time. I was ready to let the cats out freely into the garden, but Lars did not want to risk it. Maybe we could close in a part of our garden with a high fence, covering the borders with metal chicken wire buried deep in the ground, so that the cats could neither climb nor dig their way out? So we had a fence about ten feet tall and with holes no bigger than an inch built in a part of the garden between the house and the greenhouse. We told the construction workers that the fence had to be cat proof. We had the chicken wire buried about twelve to fifteen inches under the fence. Finally, the day came when we let our cats out for the first time.

They loved being outside from the very start. At first we let them out for only an hour, when we were in the garden as well. It was a pleasure to watch how they jumped around and went wild, sniffing everything, or just sat quietly in a corner. They seemed to enjoy their new outdoor territory very much.

After a few days we dared to let them out without supervision. A few times they discovered small holes in the chicken wire while playing nearby and tunneled their way out to the other side. We had to look for them for hours before we could take them home to safety again. But aside from these escape attempts (of course we always repaired the holes right away), they were good about staying in their fenced-in garden, where there was always something new for them to experience: plants, insects, birds, sun, rain, snow and occasionally a neighboring cat who showed up on the other side of the fence—they were fascinated by all of it. Then we had the idea of having a cat flap installed, so that they could get to and from the garden on their own and we did not always have to open the door for them. That did make it a bit easier.

TIP: Do you have a fat cat? Neither diets nor visits to the vet help? Try to help your cat find an exciting activity, or a hobby. Play with them every day, or give them access to new rooms or to the garden. Or try a food dispensing toy or an activity board with lots of places to hide food or treats. Try filling an empty toilet paper roll with five to ten treats, and poke small holes in the roll so that the treats can fall out when your cat plays with it. Look online or in cat books for more examples. Everything that encourages your cat to be active and makes it harder for them to get to their food has a positive effect on your cat's weight.

But how did it work for our glutton, Turbo? Wonderfully! In the meanwhile, he became much less fixated on food. Every year, when we go to the vet for his shots and checkup, he weighs a little less. Shortly after the cats went free-range, we traded Turbo's diet food for normal food, and he still did not gain weight. His new hobbies include sitting in the greenhouse and watching flies, walking around the garden, and occasionally eating a little grass. He does not run to the other bowls after every meal; instead he goes straight outside, where he sits on his favorite bench and cleans himself thoroughly.

MY CAT BITES AND SCRATCHES ME

We do not know where Vimsan came from, where she grew up or what experiences with humans she had before she came to live with us. She must have been in great pain when I found her severely injured in our basement. Still, she did not try to defend herself when I petted her, so she seemed to be used to humans. While she was convalescing, Vimsan lived in our basement so that she could have her peace and did not have to confront our other cats all the time. We visited the veterinarian frequently, changed her bandages regularly, and gave her antibiotics and painkillers, which she took bravely. I spent an hour with her every morning and every evening, and after each meal she lay in my lap, where she stayed for a long time, purring and letting herself be petted. But once, I petted her when she was still standing in front of her bowl, and she bit me immediately. I had to go to the doctor, where I was prescribed antibiotics for the deep, inflamed wound. From then on, I was a bit more careful around Vimsan.

When we introduced her to the triplets a little while later and she started to share her space with them, we tried to pick her up like we did with the others, but she did not care for it at all. She hit us with her front paws, tried to bite us with her sharp teeth and swung her tail intensely. She managed to bite me once more. We gave up our attempts to pick her up, and guessed that Vimsan's aggression might be due to bad experiences with humans earlier. Maybe she was often held against her will, and perhaps humans had also hurt her.

It occurred to me only much later that it might also have to do with the rapidity of my movements. I had already learned that two rivals who do not like each other but wanted to avoid a physical altercation removed themselves from the field with very slow movements, in slow motion. It was a signal to the opponent that they should not follow and should not launch an attack. Hence, I thought that if I wanted to show Vimsan that I am not dangerous, I should move slowly as well.

We are still working on it, but it seems to be going all right. If I approach her with my hand and pet her using slow movements, she does not defend herself and does not try to scratch or bite me either. If I touch her fur with rapid movements though, she turns around right away and hits at the air with her paw, as though to show me that she is not interested. We train every day though, and I speak softly with her, repeating the same slow movements and holding my lower arm up like a "friendly cat tail." I believe that with a little patience and a little practice, we can both learn to get along with each other. My husband, Lars, can now pick her up and put her on his lap without a problem when she rubs up against his legs. When I teach Vimsan new tricks, such as getting into the carrier by herself, I do it with a lot of patience, a lot of rewards (e.g. treats), a soft voice and very slow movements.

TIP: If your cat bites and scratches you, try to be more patient and use very slow movements when you want to pet them or pick them up. Please consult a veterinarian, cat psychologist or therapist, as there are no universally valid rules. Give your cat time to get used to your hands (or gloves) and never use your hands as a biting or scratching toy. I hope it will work as well with your cat as it did with our Vimsan.

MY CAT DOES NOT GET ALONG WITH OTHER CATS

We noticed early on that Vimsan did not like to be around other cats. If she wanted to get on one of our laps and noticed that one of the triplets was already there, she would hiss and run away quickly, as though she were afraid or angry. Then spring came and it got warmer, and Vimsan learned to climb the high fence in the garden, so that she could get out of the garden and away from our other cats. She roamed around all day and came back only to eat at night. Day after day, she climbed over the fence and ran to the neighbor's garden, where we often heard her growling, howling and snarling. When we ran to her, we usually found her in an aggressive situation with Kompis or Graywhite. At home, too, things were not going that well between Vimsan and the other three. Especially Rocky and Turbo, who liked Vimsan less and less and began to follow and sometimes even attack her.

Thus, Vimsan stayed out longer and longer. When she got home at night, she brought strange scents with her, perhaps from a plant from the neighbor's garden or from a new spot she had discovered. These strange smells made Rocky and Turbo even more suspicious. The situation escalated.

With us, too, Vimsan became increasingly reserved. She did not want to be petted and would run to her favorite spot immediately after eating—a high shelf, which she liked because she could not be ambushed there. When we went on vacation for four days, we separated our cats to be on the safe side. The triplets got the house except for the kitchen to themselves, while the cat sitter gave Vimsan her food in the kitchen. She could also spend as much time as she wanted in the garden, and come and go as she liked through the cat flap in the kitchen door.

After our return, Lars and I considered how we should best proceed. Although Vimsan had gotten along all right with the other cats for a while, their relationship had deteriorated and it was clear to us that Vimsan and the toms did not like each other anymore and fought more and more. The question was: Could Vimsan integrate into our household without forcing her or the other cats to give up their free-range lifestyle? Luckily, we have a big house. Thus, Vimsan could move into her own "suite" of three rooms. We put her favorite toys, her litter boxes (she had two: one for urine and one for stool), blankets and basket there, and gave her a safe space to eat for her alone. She could even get onto our garden fence and into the neighbor's garden through a window, so she did not even have to climb the fence anymore.

The results were apparent immediately. The door between Vimsan's "suite" and the rest of the house always stayed closed, and we noticed with relief that both Vimsan and the boys became much calmer and more relaxed. Every morning I went in to Vimsan, fed her and let her out, though she often came right back if she noticed me eating my breakfast at the same time in "her suite." She would jump on my lap, curl up cozily and stay with me purring for a long time. It was a huge difference, as she had hardly noticed us humans before. We ate our dinner with her every evening, and spent two or three hours playing or cuddling with her, which she seemed to enjoy immensely. Every day she became increasingly loving and social.

We asked ourselves whether we should not try to improve the relationship between Vimsan and the triplets after all. We still have not made a decision, as it would require a lot of work and patience from us, and the result (whether it worked or not) would remain in doubt. Some cats simply do not like other felines, and Vimsan may be one of them. We have bought a new cat flap for Vimsan, so that she can get in and out without our help. The division of rooms has become routine by now. The triplets know that I disappear into the other living area every morning and that my husband, Lars, and I do the same thing every evening. They simply go out, or lay down somewhere and sleep until we come back. For the moment, we are leaving it like that and are postponing the decision as to whether we will reintroduce Vimsan and the triplets to a later point.

Kompis initially lived exclusively in our garden, and we never let him into our house. Although we took him to the vet to treat his injuries and had him vaccinated and chipped, we did not trust ourselves to introduce him to the other cats. But then winter came, we closed the door to our hall so that the triplets could not go there, and let Kompis stay in the hall when the nights got colder. We provided him with his own food and water bowl and a soft blanket. When he had had his dinner in the hall one cool evening and was lying on his blanket, I dared to introduce him to Rocky. They had met on opposite sides of our garden fence and been lying pretty close to each other (separated by the fence) without growling or howling. Rocky—the most cautious and anxious cat in our family—had already seen Kompis through the window in the door, and was happy when I opened the door. He hurried toward Kompis with a friendly raised tail. I held my breath. Would Rocky and Kompis become friends? But when Kompis saw Rocky, he was scared and went to hide under a chair. Of course I carried Rocky right out again, then went straight back to Kompis and promised him that he need not meet any other cats if he did not want to.

TIP: If your cats do not get along with each other, try separating them for a time and see if things improve. After a few weeks or months, you can try reintroducing them. Please consult a veterinarian, cat psychologist or therapist first. Many experts recommend proceeding in this order: First use scent (trade blankets and baskets, and try to produce a shared group smell), then sound (first through a closed door, then through a door that is slightly ajar), then sight (it is best to use a screen door, so that the cats can see, hear and smell each other, but cannot attack). You can then try to elicit good experiences by giving them food, treats and toys when they sit peacefully across from each other at the screen door without demonstrating aggressive behavior. If everything goes well, you can open the screen door under your supervision for a little while every day. But please be patient and be prepared for the possibility that it might not go well.

And that is how it still is today. Kompis has his own "suite" in our relatively large hall with the guest bathroom and often sleeps on his favorite footstool. We have now installed a cat flap for him as well. It responds to the code in his chip so that he can come in and out at any time, but no unfamiliar cats can come in. When he comes home in the evening, Lars and I often sit with him in the hall for a while and talk to him. Kompis is a very social cat, who likes spending time with us, both in the garden and in the house.

MY CAT PEES EVERYWHERE

When Vimsan was still living with the others, we noticed on multiple occasions that one of them had missed the target while peeing. At first we suspected Vimsan, as she sometimes peed in the sink in our bathroom or in the kitchen. We assumed that she did so because her former people had not cleaned her litter box often enough, or perhaps did not even have one for her, so that she was forced to pee in the sink. As long as it went into the sink it wasn't too bad. But when we found puddles on our carpets, on clean laundry, in the cat's baskets and even on the kitchen stove, and when we had to search more and more often (once we even searched for days until we found a puddle in a plastic bag on a shelf), we started to suspect the other cats.

Adding to our suspicions was the fact that Vimsan hardly had any reason at all to pee in the house anymore, as she was out and about almost all day and she could do her business in the neighbor's garden. And even before Vimsan came to us, we had found cat poo on the carpet in the playroom a few times. We had assumed at the time that it was from the anxious Rocky, who might have been scared by a noise on the street and might have "let loose" as a result.

Confronted with this problem, we went searching for tricks and tips that might help us deal with this problem. We found them online and in books. We read a lot about why cats become unclean and tried a lot of helpful tips. We bought new litter boxes (so that we had six litter boxes for four cats), and tried other kinds of kitty litter; we placed the litter boxes at different places in the house (two on the ground floor and three more upstairs, well distributed, so that they could do their business undisturbed, and then an additional litter box on the spot of the last puddle). We took all four cats to the vet to see if a bladder infection could be the cause for all the peeing outside the litter boxes. At night, we separated all the cats to see where the new puddles would be the next morning. We caught Rocky in the act once. He had just tinkled on the entry mat by the entry door and was busy scratching away, trying to hide the puddle.

We took Rocky to the vet again, who said he was 99 percent certain that it was a behavioral problem. Rocky was marking his territory with urine because he did not want Vimsan around him. The vet suggested finding Vimsan a new home. But we loved all the cats equally (even if they did miss the mark when they peed), so we started to look for other solutions. Should Vimsan move back into the cellar? But we seldom spent time in our cellar, so wouldn't Vimsan feel lonely? Moreover, we would never know whether she was down there or outside in the garden without going downstairs to check on her. We were at a loss. The peeing problem had now gone from bad to worse. We rarely found puddles in the same place; they were always someplace new. Every morning and every evening we went sniffing through the house trying to find out where a puddle might be hiding. It was a tough time and when I discovered nine (!) new pee stains one Saturday, I had had enough. Something had to be done!

This all took place at the same time as the events depicted in the last section. As soon as we set up the three-room "suite" for Vimsan and noticed that the triplets grew calmer, the errant peeing miraculously stopped as well. For a while, we still found a couple of old stains, but after a while we didn't find any at all. Success! And although Vimsan did not live with the other cats anymore, all of the cats were much happier—and so were we!

As already mentioned, we are still waiting before we decide if we want to bring the cats together again. But if we do, we have to be very careful and perhaps ask a cat behavior expert for help. We might give clicker training a try. (I got this tip from the German cat expert and behavioral consultant Birga Dexel [2014] at a cat conference in Austria. She has had a lot of positive experiences with clicker training.) This method includes conditioning the cat to a certain sound produced using a clicker and followed by giving the cat a reward (a treat, cuddle or toy). In this way cats may learn new things, for example, that it is actually really nice when everyone lives together.

HELP—THERE IS A STRANGE CAT IN MY GARDEN

There were also problems that we could not solve. We could not save the beautiful but very sick cat Red, who came to our garden and was fed by us every day for several years, because we waited too long. If we had taken him to the vet sooner and contacted the humane society he might still be alive, enjoying a nice life in the country. Why did we think for such a long time that he had a home where someone cared for him? Is our human perception so limited that we do not notice when a cat is not doing well?

After the experience with Red, I often thought about how we might become more conscious of cats in our neighborhood who were sick or had accidently run away from home and could not find their way home, or wandered around homeless. And what can we do to help these cats? More on that in a second.

Another problem arises when a cat owner cannot keep their cat for some reason. Maybe somebody in the family became allergic, or the cat owner has to move into an apartment where cats are not allowed. Other reasons might include the human falling ill, or taking a new job that makes keeping a cat impossible for lack of time.

There are many books and online sources where one can find advice on this topic. I have acquired some valuable information there and have been able to help cats a couple times, even though I could not take them on myself. We have reached our upper limit with five cats; and although we have divided our house into three cat "suites," we do not dare to take on more cats. We do not have time for more cats (and more cat problems). I have discovered that a little often goes a long way for a cat though.

Cats have a low status in our society. You can often get a kitten for free (a dog, in contrast, costs a pretty penny). As a result, many people do not think it through enough before getting a cat, and know little about their needs. Unfortunately, many cat owners do not vaccinate, fix and chip their cats. Unchipped cats are difficult to reunite with their humans if they run away from home. Cats who are not spayed or neutered—whether they are homeless or family pets—quickly lead to even more homeless cats.

Many thanks are owed to every animal sanctuary and humane society. They have a very important job, as, due to the low status of the cat, many people think that it is not a big deal to put an animal who is not young and sweet anymore on the doorstep. They assume it will somehow be okay on its own. But this is not true! Especially not if the animal is used to living with humans. Every single cat needs food, water and a warm, dry home every day, one where they feel safe and well. They have to be taken to the vet for vaccinations and checkups and when they get sick, they need someone who takes care of them and treats them until they are well again.

Most animal sanctuaries are overflowing and cannot

take on any new guests, but they often have waiting lists and if it becomes impossible to keep a cat, one can certainly find a spot on a waiting list. My husband and I regularly make donations to cat sanctuaries, whether through a bank transfer or by tossing something in the tin at the supermarket. We have seen with our own eyes that the animal sanctuary can truly mean the difference between life and death for many animals. Please never simply set an animal on the doorstep. Find an alternative residence in a sanctuary until a new home can be found. You will probably be saving its life.

When we had been living with our triplets for a few months already, a pretty, fixed gray tomcat came to our garden and asked us for food. He seemed to be accustomed to people and he used to come and finish Graywhite's breakfast. Of course he got extra food and water from us, as we knew that one should give a strange and hungry cat something to eat before contacting the local authorities. It was summer and very warm. "Little Gray," as we called him, slept on a blanket in a party tent that we had in our garden. He was very affectionate and often came to lie in our laps. We dewormed him and treated him for ticks and for fleas, so he would not infect our cats. It so happened that a woman from the humane society came by shortly afterward to see how Donna, Rocky and Turbo were doing, and I asked for her advice. The police were not aware of any cat owner who had registered a gray tomcat as missing, and no one had responded to the posters that we had hung up in the area about a found cat.

It turned out that the chairwoman of the local humane society was going to give an interview to the local paper shortly, so she suggested that the chairwoman say something about Little Gray there. The article appeared the next day, and the chairwoman got in touch with us after a couple of days and said she might have found Little Gray's owner. When he came to our garden with his carrier, Little Gray recognized him right away, and it was a happy reunion.

The owner, who lived alone with Little Gray, got sick and had to go to the hospital for two weeks. When the cat sitter visited to feed his cat, Little Gray slunk away unnoticed and could not get back in, as the window that was his usual entryway had been closed by the sitter. So Little Gray had no choice but to find something to eat and somewhere to sleep elsewhere. Well, he had found the right place. We were so happy that we had been able to help the friendly and pretty Little Gray reunite with his owner.

Last summer, there was a heat wave in August and we often sat in the garden until late in the evening, where Kompis usually kept us company. He lay in our laps, slept on his blanket, or played boccie with us or with twigs and blades of grass, which he liked to chase as we moved them about. One night, we noticed that he kept looking into our large rhododendron bushes. As I looked into one bush more closely, I saw a long-haired gray cat with a white nose and chest. At first, we thought that a new neighbor cat had moved in somewhere close, and left it alone.

However, the next night we found him in the rhododendron bushes again, and the night after that, too. He (we assumed it was a he, because he was pretty big) and Kompis observed each other and howled at each other on occasion. Whenever we put out food for him, he ate it immediately and asked for more. Although he was afraid of Kompis, he occasionally ventured from his rhododendron and allowed us to pet him and hold him on our laps. His fur was tangled, but he seemed to have had a good home once. We hung up signs at the vets and in stores. Because the newcomer was a pedigree cat, we took him to the vet to see whether he had been chipped. There we learned that there was already a notice of a missing cat that matched our foundling. The posting came from a cat owner who had accidentally left a window open, allowing the cat to escape. Because he was not an outdoor cat, he grew afraid, ran away and got lost. The owner was in despair, as the cat had already been missing for more than a week.

That night, we set up a nice room for the long-haired gray cat in our basement, complete with litter box, water and food and drove directly to the vet with him the next morning. "I will call the owner right up and ask if he can come," said the veterinary nurse. The owner came a few minutes later, recognized his cat right away and was very happy. I lent him our carrier so that he could take his cat (who was called "Kis," by the way) home with him, and when his wife and daughter returned the carrier an hour later, they brought a large box of chocolates with them as a thank-you. We were really happy that the cat had found his way back home...and Kompis seemed very happy to have the garden to himself again.

Last fall, another strange cat appeared in our garden. This time, it was a young brown-and-white tomcat who had not been neutered. He and Kompis did not get along at all. There were repeated physical altercations. The two of them howled and growled so loudly and for so long that we often had to go outside and bring Kompis inside so that we could sleep. Again, we assumed initially that it was a new neighbor cat, as he was often away for two or three days, but then he suddenly returned and seemed to be really hungry.

We hoped that his owner would take him to the vet

to be neutered soon and that the fighting would then stop, as we were growing worried about Kompis, who often returned home with scratches and even bite marks on his face. It was clear that he didn't like this new cat at all. A few times we even chased him from our garden when he was being especially rough with Kompis. I had a terribly bad conscience afterward though, as I had the foreboding that the cat might not have a home after all. I started to observe him more carefully and noticed that he regularly emptied Kompis's food bowl, which we left in a little hut in the garden. I also noticed that his fur was not so shiny, and that he occasionally slept on Kompis's blanket outside if Kompis was inside for the night.

We hung up notices and informed the police one more time. At the same time, we considered whether we should try taking on one more cat after all. But we had enough to do with our own cats' problems and realized that it would not be a good idea to add one more, especially as the new one was thrashing Vimsan as well as Kompis.

Finally, I wrote an email to a cat sanctuary near us and asked if they could put the tomcat on their waiting list. I got a reply with good news the very next day. "We have found homes for many of our cats recently, so we have a spot right away. When can you bring him?" It was negative sixteen degrees Celsius (about three degrees Fahrenheit) that night, so we did not have a problem coaxing the new cat into our basement with some food. I was allowed to pet him and sat with him for the whole evening, happy that he would have a warm new home and would not have to spend the entire cold winter outside in our garden.

We took him to the cat shelter the next morning. It was the first time I had visited such a place, although I had seen many on television. It was a very beautiful and well-maintained shelter that provided a large room for each cat, and it was warm. The director had prepared the cat's room with warm blankets and three different kinds of food: "Just in case, as we don't know what food he prefers yet." When I let him out of the carrier into his room, she approached him and was allowed to pet him. She said, "What a beautiful and nice cat. I don't think we'll have any problems finding him a home." Working with the director of the sanctuary, we placed an ad on the website of the cat sanctuary, and offered to pay for his costs (vaccinations, neutering) at the vet's. Just two months later, we learned that "Teddy" (as he was called from that day forward) had found a new home. We continue to think about Teddy regularly and are glad that it ended well for him.

So what can you do when you find a cat that seems to be homeless? I would start by finding out whether it is hungry. I would then feed and water the cat whenever it comes by, but leave food only if I were sure it is not a neighbor cat who is only visiting your yard. In winter (if the temperature is below freezing), I would add a little cooking oil to the water so that it does not freeze right away. If it is a healthy, well-maintained cat, you can observe it for a while and leave it outside, so long as the environment is safe. Kittens should be brought inside only once one has looked carefully for their mother. One should never separate a mother from her young. Injured or sick cats should be brought to a vet as quickly as possible; healthy cats should be offered a spot where they are protected from the wind and the weather, as well as a warm blanket. I would also try to have as much contact as possible with the cat. I would see how it responds to humans, whether it is in pain, is pregnant or shows signs of unusual behavior. I recommend keeping each new cat separate from your resident cats, so as not to risk spreading diseases.

Then consider how you can best help the foundling. I would go online and search for a website with information about regional resources for helping cats. What local authority can one call? What humane societies or animal sanctuaries are there nearby?

Create a flyer with a photo and hang it in the area around your home, as well as in nearby stores. It is advisable to require someone claiming to be the pet's owner to provide a photo of them with their pet to ensure they are legitimate. Bring the cat to a vet to check whether it has been chipped. If the cat has a collar, check for the telephone number or address of the owner. I would buy a collar for a cat who does not have one, and add a little note with your cell number: "Is this your cat? Please contact..." If no one gets in touch, put the cat on the waiting list of an animal sanctuary and help with food, water, company and protection until it finds a place.

If you give a homeless cat just a little of your time and patience, you can help it to find a new and better home. It is not particularly difficult, and you just might save their life.

You are approaching the end of this book. I hope you enjoyed reading it and maybe learned something you didn't already know about cat communication.

I certainly had a lot of fun writing the book. I laughed—and cried—as I relived my own experiences with cats by putting them into writing. Cats are wonderful animals, and our relationship to them can only improve by learning to better communicate with them.

12

STUDIES AND PROJECTS

Although this book is not an academic thesis, and is not intended to be one, I cannot help myself from sharing some of the results of my earlier research on cat sounds. You will see that it really is not so hard to conduct a small study. A phonetician's best tools are their own ears, followed by an audio recorder (a smartphone or a video camera which allows you to record sound, for example), which make it possible to repeatedly and carefully listen to the same sound. That alone is enough to set you on the trails of some of the mysteries of the language of cats. You can learn a lot about some of the phonetic characteristics of the various sounds and their possible meanings this way.

In this chapter, I will present my previous phonetic studies to you. I conducted them in my free time, in the evenings and on the weekends, as during the day I am busy with my job as a teacher and researcher in phonetics at Lund University. My primary goals were to satisfy my own curiosity and to learn more about these cat sounds, which were so mystical to me back when I started.

With my descriptions of the studies, I would like to inspire you to try doing some research yourself—everyone can investigate cat sounds using simple methods and a little interest, time and patience. If it is too academic or too boring for you, no problem. You can easily skip this chapter and still understand the rest of the book.

MY FIRST STUDY: PHONETIC CHARACTERISTICS OF PURRING

I already mentioned the conference in 2010 where Dr. Robert Eklund held a lecture on his studies comparing the purring of a house cat to the purring of a cheetah, and discovered a great number of phonetic similarities. His lecture made me realize for the first time that maybe I could also contribute something to the research on cat sounds.

Once I got home again, I got out my old video camera and recorded the purring of our cat Vincent. It is not entirely straightforward to record a purring cat. I had my video camera on constant standby, and when Vincent was lying in his favorite spot resting, I tiptoed up to him with my camera, pressed the record button and petted him gently and carefully until he began to purr. Then I carefully put my hand on his body so that I could feel the movement of his breath.

Whenever his body lifted with an inhalation, I said "up" and "in" loudly into the microphone, and when it sank again as he exhaled I said "down" and "out." I did that so as to distinguish the exhalation phases from the inhalation phases. Then I filmed for approximately another minute without commentary, so that I could record a sufficient number of inhalation and exhalation phases as materials for my research.

When we had to put Vincent to sleep and the mischievous triplets moved in with us just a few months later, I recorded their purring with the same methods. After that, I analyzed the sound recordings together with Dr. Robert Eklund using acoustic-phonetic methods. Using a computer program for speech analysis called Praat (it can be downloaded free of charge at www.praat.org) we measured the length (duration), volume (acoustic sound level pressure or intensity) and pitch (melody) during both inhalations and exhalations and compared the results of all four cats.

The following figure shows our analysis in Praat. In

this example, we can see that the intensity (loudness or volume) and fundamental frequency (pitch contour or melody; the curve in the third pane) are higher during exhalations than while inhaling (which can be recognized by the greater amplitude in the upper two panes and the higher contour in the bottom pane). Inhalations (or the ingressive phases) are marked with an *I*, and exhalations (or the egressive phases) are marked with an *E*.

The two upper panes show the waveform of the microphone signal (the top one is filtered to show only the lower frequencies, and the second is the original signal), the third is a depiction of the frequency analysis (a spectrogram) with a fundamental frequency contour (bottom curve), while the bottom pane shows the division of the audio recording into ingressive phases (inhalations) and egressive phases (exhalations).

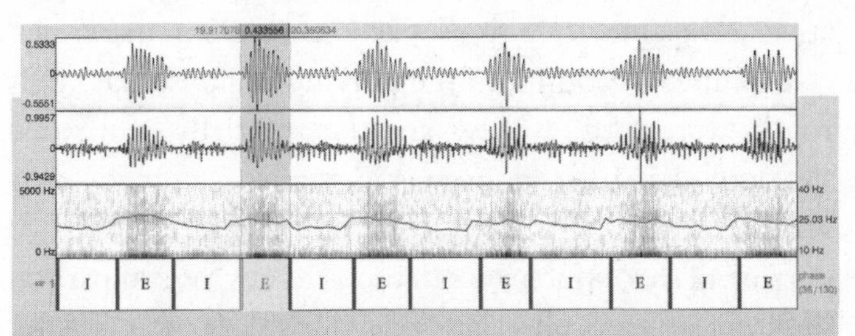

Acoustic analysis of purring with the speech analysis program Praat.

The results of the study showed that two of the cats purred significantly louder during the egressive phase (exhalation) than during the ingressive phase (inhalation), while there were no differences in volume between inhalations and exhalations in the other two cats. The length of phases varied significantly between cats, but all cats had significantly longer inhalations than exhalations.

The fundamental frequency of vibrations is very deep in purring (roughly between 21 and 27 Hz), and all four cats were in the same frequency range. This is in line with other studies of feline purring. Two of the animals had a significantly higher fundamental frequency during the egressive phase, while it was the other way around with one cat, and there was no real difference between inhalations and exhalations with the fourth.

As far as I am aware, our study is the very first comparative and quantitative (it uses acoustic measurements and results) investigation into the purring of house cats. Our results were partially new, and they partially confirmed the results of earlier (nonacoustic) studies. The study provides an acoustic reference point for the peaceful, satisfied purring of cats, which we can use as a basis for comparison in future investigations into other forms of purring. After all, there are a few other situations in which cats purr.

A few years ago, a team of researchers in England discovered a "cry" embedded in purring (see page 127). They traced it back to the fact that cats purr much more loudly, or even scream, when they really want something (to get their human out of bed so they can have breakfast, for example). Further studies could therefore compare other types of purring in order to determine whether the pitch, volume or length of phases is different in other kinds of purring than in the results of the investigation above.

TIP: How does your cat purr? When does it purr? Is it possible that you can even distinguish between the purring your cat does while it is relaxing and the purr that it uses as a greeting when you get home from work? Maybe you can make a contribution to the research into cat sounds by recording different variations of purring and investigating them using phonetic methods, among which the first is careful listening.

MY SECOND STUDY: FRIENDLY CAT SOUNDS DIRECTED AT HUMANS AS WELL AS OTHER CATS

After completing my study on purring, I simply could not stop listening to my cats with my "phonetic ears." My curiosity had been awoken. I wanted to collect detailed information about other cat sounds so as to be able to systematize them. Thus, I started following our triplets with my video camera, recording many different situations in which they made every possible sound. After a month, I had recorded 538 sounds, and I was quite proud of this myself, as it is not easy to record a cat at the exact second when they happen to say something. You first have to observe the situations where most sounds occur and be ready with a camera and a microphone in these moments. In our house, it was mostly when the cats were being fed or getting a treat, when one of the cats wanted to play with us humans or with its siblings, when one of the cats came up to us and we greeted it with affection, or when a bird or insect caught the attention of one of the cats.

I investigated these 538 sounds carefully and tried to subdivide each sound into one of five relatively rough categories: meowing, trilling, trill-meowing, chirping and other sounds. The following table shows the number of sounds per category that I had recorded from each of the cats as well as the total number of recorded sounds. The study focused only on friendly cat sounds, so I was investigating only some of the categories that I have described in the book until now. It was one of my first studies, and I did not yet know exactly how many and which categories there might be. In the category "other sounds" I put all the sounds that did not occur often, in this case purring and longer phrases with several different sounds. I did not continue investigating this category in this study, as it contained only eleven sounds—too few to be able to draw any general conclusions about them.

Cat	Chirp	Meow	Trill	Trill-Meow	Other	Total
Donna	1	21	18	29	4	73
Rocky	14	22	63	52	1	152
Turbo	3	36	103	165	6	313
Total	18	79	184	246	11	538

Number of sounds per cat recorded over a month, listed by category.

One can learn a lot from a table like that. For example, I saw immediately that Turbo was the big chatterbox and that Donna had not said much while I was recording. So Donna had produced the fewest sounds (73), Rocky somewhat more (152) and Turbo the most (313). The most common sound was the friendly and inviting trill-meow (246), followed by the greeting sound of trilling (184), then meowing (79) and chirping (18).

When I organized the sounds into categories, I did so according to the following criteria: length in seconds and the fundamental frequency, especially the lowest (minimum), the highest (maximum) and the average (mean). What all the sounds had in common was a very large variation in melody, often between 100 and 1000 Hz. It is a much wider range than we humans normally have in our voices. Meowing had the highest average fundamental frequency at 698 Hz, while trilling had the lowest at 358 Hz. All sounds aside from the trill-meow were also frequently of similar length. Length was measured at approximately a half second. Trill-meowing was significantly longer (approximately 0.8 seconds), which can be attributed to the fact that it is a complex or combination sound.

These results showed that there was a wide frequency range in the melody within each sound category, much larger than I had imagined. They also demonstrated that our female, Donna, had a much more high-pitched voice than her brothers, which can be attributed to the fact that Donna, as a female, is smaller than her brothers, Rocky and Turbo, so she has a smaller larynx (voice box) and a smaller mouth. With us humans, too, women and children generally have more high-pitched and brighter (with acoustically high resonances) voices than men.

As my study included only some 500 sounds from three cats, I could not draw any far-reaching conclusions from my results. In order to make general statements about cat sounds, one naturally needs a great number of additional recordings of cat sounds. So that was my next goal.

This study provided an additional insight. Perhaps there was something wrong with Donna, who had produced so few sounds. I continued to observe my cats and noticed that the boys often pushed themselves to the front and claimed a large part of my attention, while Donna stayed in the background. Had I been neglecting her? Did she need more time with me, so that she would show herself and become more talkative? So I started spending more time with her. I played with her, talked to her and paid attention to her, especially when Rocky and Turbo were nearby. After a few weeks, I noticed a big difference. Donna did become more outgoing and produced significantly more sounds.

TIP: You, too, can produce a similar study about the sounds of your cat. Record them for an hour in situations where they normally communicate with sounds (when they are hungry, want to be let out, or want to say hello to other family members, human or animals). Do this over a set period of time—a week, a month, every Saturday until Christmas, or whatever works for you. You certainly know when and under what circumstances your cat "speaks." Then listen to the recordings repeatedly and carefully. To which category does each sound belong? Count the sounds in each category and create a table, which will make the comparison easier and tell you which sounds are the most common and which sounds are the rarest in your home.

MY THIRD STUDY: CHIRPING AND CHATTERING

I had not yet found anything about the chirping sounds that I had recorded in my second study in any academic journal article or book. I could find only descriptions of such and similar sounds online, on websites devoted to cats and their behavior. I grew curious and wanted to learn more about these mystical sounds.

So the next winter I arranged for a veritable bird buffet just outside our kitchen window in our garden, with bird balls, sunflower seeds, apples and peanuts. The birds quickly discovered the feast and the cats naturally noticed the birds on the other side of the window and made themselves comfortable on the windowsill, so as to be able to observe better. I put my old video camera in position and sat on the couch in the adjacent living room, remote control at the ready. In this way I could make video recordings of my cats without disturbing them. Every time I saw that one of my furry roommates was sitting on the windowsill watching the birds, I recorded their chirping and chattering sounds with the help of the remote control. After three months I had collected 255 sounds and began to categorize them.

It was not as easy as I had thought it would be to categorize these truly strange sounds. Because there were relatively few descriptions of these sounds in the research literature, it was difficult to find the right name for each sound. "Chattering" seemed to be the general term for all of these sounds, but because there were a great number of different variations, it seemed suitable to organize them in subcategories with appropriate names.

I did so by looking in the dictionary for names for bird sounds. Then I chose the name for each category that best corresponded to the cat sound. "Chattering" was often described as sound produced with the lower jaw and/or teeth chattering. That is why I use the term "chattering" only for the subcategory of voiceless chattering of the teeth. Next, I selected suitable names for the other subcategories. For example, *Webster's New World Dictionary* describes a "chirp" as "a short, high-pitched sound, such as the one a bird or small insect makes," which seemed to correspond best to the typical *aehh, ehh, ehh* cat sound, as my cats often produced this sound with very high-pitched and bright voices. The softer variant of *aehh, ehh, ehh* sounds more like *uyh* or *hew* and was given the name of "tweeting." Longer tweeting sounds with more varying melody were given the name "tweedling" or "warbling." The following table shows the distribution of the sounds by category for each of my cats. Again, Donna was the quietest of the bird watchers, while her brothers "spoke" more to the birds. Chirping was the most common sound, with 169 examples, and the voiceless chatter was the least common, with 22 recorded sounds.

Cat	Chatter	Chirp	Tweet	Tweedle	Total
Donna	3	19	6	6	34
Rocky	7	70	19	22	118
Turbo	12	80	9	2	103
Total	22	169	34	30	255

Number of recorded sounds of three cats over a three-month period in different chirping and chattering categories.

After I had categorized all 255 sounds, I could start my measurements of length and pitch. The individual chattering sounds were the shortest (roughly 0.03 seconds), chirping and tweeting were also very short (roughly between 0.15–0.20 seconds), while tweedling was the longest (approximately 0.5 seconds). The chattering sounds were all voiceless, but chirping, tweeting and tweedling all averaged around 600 Hz. Despite the short length I discovered a wide pitch range in these sounds (between 230 and 1200 Hz for chirping). These results demonstrated to me that cats can vary their vocalizations by altering the pitch and melody. I did not yet know whether this was instinct or learned behavior. Maybe further studies would help me to discover more about these variations in melody.

TIP: What sounds does your cat make when it sees a bird or an insect? Some cats only chatter, while others tend to chirp and some "speak" to the small prey animals on the other side of the glass with tweeting tones. There are also cats who produce a wide variety of chirping and chattering sounds. If you are not exactly sure, put some bird food outside the window and listen carefully to how your cat "speaks" to the birds while watching them. Or make a video recording and listen to the sounds carefully and repeatedly. Even if you do not have any desire to turn your listening into a little academic research project, you will have a lot of fun listening, as they are often strange and funny sounds.

MY FOURTH STUDY: THE PHONETIC DIFFERENCES BETWEEN HAPPY AND SAD MEOW SOUNDS

Early on it became clear to me that the meow sounds my cats made differed a lot depending on the situation or context. When I took them to the vet's and they sat in their carriers in the waiting room looking at the other humans and animals anxiously, the sounds they produced were entirely different than when they were at home hungry in the kitchen and saw that I was preparing their food or a treat. Why was that? And did only I hear this difference, because I know my cats so well (or perceive sounds more exactly with my phonetic ears), or do others also hear the same differences?

In order to investigate, I worked with Dr. Joost van de Weijer, also a linguist and a cat fancier, to conduct an experiment in which thirty human listeners were asked to judge a number of meow sounds: six sounds that were recorded in a feeding context, and six meow sounds that were recorded in a veterinarian context, arranged randomly. After each sound, we asked the listener if the sound was from a feeding situation or from a veterinarian situation. It actually could be determined that there was a different melody in different situations. The feeding-related meow sounds varied more in the melody and more often ended with a rising tone (intonation, melody) than the veterinarian-related meow sounds, which were more likely to end in a falling tone and displayed much less variation in the melody. The following diagram shows the melody of the feeding meows (above) and the veterinarian meows (below).

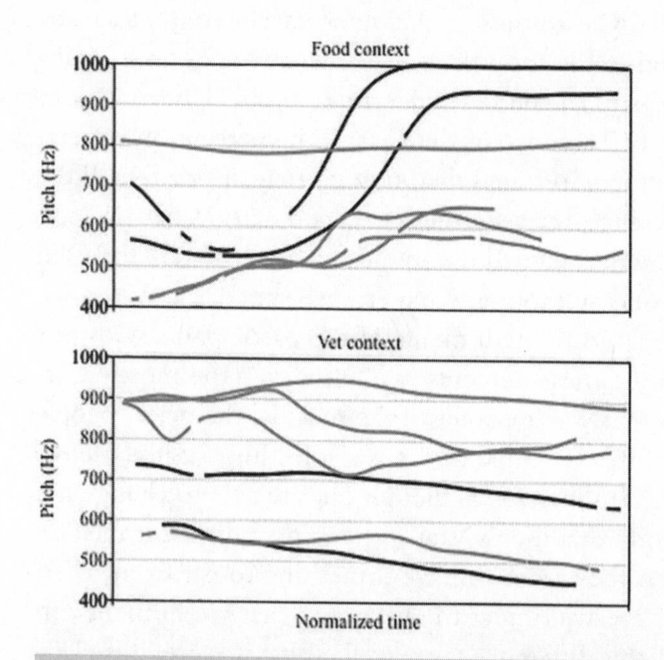

Melody of meows in the feeding context (above) and in the veterinarian context (below). The black curves show the sounds that were easiest to determine in which context they were recorded.

You can listen to the examples on my website under the category "Meowing Sounds." Cat sounds in the feeding context can be heard in the sound category "Meowing Sounds," subcategory "Meowing (Or Miaowing)," under the keywords "Donna and Turbo meow to solicit food." Cat sounds in the veterinarian context can be found in the sound category "Meowing Sounds," subcategory "Moaning" under the keyword "Donna, Rocky and Turbo moan at the vet's." The perception experiment demonstrated that most participants could tell the difference between these two contexts. What is more, cat owners judged the sounds more accurately than did listeners without any experience with cats. When we compared the judgments of the participants with our acoustic measurements, we found that the more variation that was apparent in the melody, the better people were able to judge that it was a feeding-related meow. It therefore seems as though cats are able to change the melody of a meow, thus producing a different melody when they want to have something to eat from when they are waiting at the vet's, and that we humans can hear this difference fairly well. What is more, it is plausible that cat owners can hear the difference more easily and accurately.

TIP: Try it yourself—listen to a cat sound you have recorded yourself repeatedly and carefully. Concentrate especially on the melody. Pay special attention to whether the tone is rising, falling or even. The longer the sound, the easier it is to hear the melody. As a next step, you can record sounds in two different situations and listen to them one after the other. Can you hear differences in the melody? If so, what do the differences consist of? Are they at the beginning or at the end of the sound? Is the pitch in one sound deeper than in the other? Are there other differences? This is an easy way to investigate the melodies of the sounds. If you also know the situations in which the various melodies occur, you can understand the "language" of your cat a bit better.

MY FIFTH STUDY: EMOTIONS AND INTONATION IN CAT SOUNDS

It is often difficult, or even impossible, for me to judge the emotional state of my cats. Are they happy and satisfied, or simply tired? Are they excited because they are happy about something or because something is annoying them? We cannot ask our cat what is wrong, or how they feel in a particular moment or situation.

But maybe, by investigating the intonation or melody and other characteristics of their voices, we can learn something more about the emotions of cats. In a small experiment, I asked thirty-six participants to judge the emotions in twenty-eight cat sounds. The participants were students of phonetics, and the various sounds were from my own as well as other cats that I had recorded in various situations. Because I knew the cats well and also knew what happened before and after each sound was produced, I attempted to identify the feelings that most closely corresponded to each sound. As feelings are seldom easy to evaluate, I made it a bit easier for the participants by asking them to choose an emotion from one of the seven following categories:

1. Joy: happy or content
2. Sorrow/Fear: sad or afraid
3. Anger: angry or discontent
4. Desire/Need: questioning, begging, wanting or hungry
5. Neutral (no special feelings)
6. Other (a feeling that does not belong to any of the categories above)
7. No idea (only to be chosen if none of the other categories apply)

In addition, I asked the participants to rank their experience with cats, as well as the difficulty of the whole task, on a scale from one to seven.

After the perception experiment, I asked my participants which clues or hints they used in their judgments so that they could assign the sounds to categories. Many people said that they had associated sounds with a low average pitch and a high degree of noise as Anger, and sounds with a high average pitch and a low degree of noise as Joy. Some associated Desire/Need sounds with a rising intonation and Sorrow/Fear sounds with a falling melody. I used this information when I examined the cat sounds more precisely. What is more, the descriptions provided by my participants seemed familiar to me somehow. They somehow corresponded to descriptions of human feelings as well as to Ohala's frequency code. Perhaps that is one reason why humans and cats understand each other reasonably well.

The results showed that, of a total of 980 estimations provided by the participants, 350 (36 percent) were correct. The most correct evaluations were of the two purring sounds (Joy). A trill as well as a trill-meow with a low pitch were the most frequently misinterpreted. A possible explanation for these results could be that the participants had very different degrees of experience with cats. Purring—next to meowing—is the sound most often associated with cats, but trilling, especially in its more low-pitched and darker (with acoustically low resonances) variation (grunting), is not as well-known and could be evaluated as an aggressive sound by participants without any previous experience with cats. However, since low-pitched and dark trilling is used as a friendly greeting sound, this could lead to miscommunications between cats and inexperienced humans. The pitch and the melody of cat sounds are, perhaps, more complicated than the frequency code, and a precise, systematic investigation into the melody of cat sounds could certainly bring more to light about the meanings of various cat sounds, improving communication between cats and humans in the process.

TIP: Maybe you can make your own contribution. Try to recognize the differences in your cat's sounds (voice register, voice quality, melody and other elements, like noise or frication) and describe the various situations in which your cat demonstrates emotions. Is the register higher (or brighter) if your cat is happy or excited? Is there more noise in the sounds used when your cat expresses its discontent? I also continue to investigate the sounds of my cats and try to associate them with various emotional states.

When I started this study about emotion in cat sounds, I was already prepared for the possibility that it would not be simple for either me as the researcher or for the participants. Many sounds were evaluated differently by me and by the participants, and there were also deviations among the participants. So the study did not really produce any results. I remain convinced, however, that a more precise study of the feelings signaled in cat sounds would help us to better understand the language of cats.

MY SIXTH STUDY: AGGRESSIVE AND DEFENSIVE SOUNDS IN CATS

When Vimsan came to us injured in the fall of 2014 and we made her a part of our family, our triplets did not like it at all. Many of my video recordings show conflicts between our four-legged roommates: Vimsan trying unsuccessfully to make friends with Donna, and Donna howling, growling or hissing in return; Vimsan walking past Rocky and Turbo, and the boys following her with growling noises and evil stares. It provided me with material for my research into aggressive cat sounds. I tried to figure out how many different sounds cats employ in agonistic situations, and how they can be divided into subcategories. What do they sound like, and what are their phonetic characteristics? Are these sounds all used in the same situations or does hissing belong to one category of aggression and growling to another? In order to learn more about the unfriendly sounds of cats, I followed little Vimsan around with a video recorder for eight days. I wanted to know which sounds occur most frequently, which phonetic characteristics (pitch, melody, voice, elements of noise) can be attributed to them, and whether they occur in the same aggressive, combative, defensive situations.

This was the first time I realized that being a researcher and being a cat owner might come into conflict with one another. As a researcher, I want to record as many different kinds of cat sounds as possible, so that I can study them more closely, while as a cat owner and fancier, I am more concerned that my darlings are not put in unpleasant situations too often. But I need such situations if I am going to record aggressive sounds. At any rate, I have always tried my very best to make sure that my cats were not injured or otherwise harmed during my research project.

The experience taught me that my work as a cat researcher requires a delicate touch. If I judge that a situation is not all that dangerous, I can use it to capture a few minutes of recordings on my video camera before I intervene. Cats often put themselves in dangerous situations, but I certainly never encouraged them to put themselves in such situations. If it does happen and I am there with my camera, I stay alert; I have learned if I produce loud and catlike warning sounds (snarling and hissing seem to work best), I can often help to deflate tensions between two cats. Before I could deal with fighting situations very well, I often intervened too early in potentially dangerous situations and missed out on lots of interesting cat sounds.

After eight days, I had recorded 468 sounds and could begin sorting the sounds into categories. I identified six different categories.

Cat	Growling	Howling	Howl-Growl	Hissing	Snarling	Spitting	Total
Donna	13	175	114	38	3	22	365
Rocky	2	47	1	4	0	2	56
Turbo	13	2	4	7	0	1	27
Vimsan	3	6	0	5	4	2	20
Total	31	230	119	54	7	27	468

Aggressive and defensive (agonistic) sounds of our four cats recorded over eight days, divided into six different categories.

In this investigation, Donna was the "chatterbox," with 365 out of 468 sounds. Rocky and Turbo were significantly quieter, with only 56 (Rocky) and 27 (Turbo) agonistic sounds produced. Our new cat Vimsan was the quietest, with only 20 sounds. Howling was the most common sound (230), followed by combinations of howling and growling (119). Growling without elements of howling was rather rare (31). My cats hissed 54 times, spat 27 times and snarled only 7 times.

My cats used growling, howling, and a combination of

howling and growling as warning sounds when another cat got too close. In general the sounds were between two and four seconds long—so relatively long. Hissing, spitting and snarling were much shorter (around half a second). Hissing and spitting were also used as a warning, and spitting was often combined with a jump forward and a little stamping of the front paws.

During these eight days, there was only one physical altercation between Donna and Vimsan. It was very short and explosive, and then the two cats jumped apart and started to howl and growl again. So far, I have only recorded snarling with my video camera just before or during physical altercations between cats. After eight days, our cats gradually got to know each other better, and the aggressive situations grew rarer. Vimsan had become a part of our family.

TIP: If you are thinking about welcoming a new cat into your family, inform yourself carefully about the best ways to introduce a new roommate to your resident feline companions. Arm yourself with a large portion of patience and please do not think that everything's going to go off without a hitch. Observe the sounds and the visual signals the cats give off when they meet for the first time and try to interpret them before it comes to physical violence.

13

CURRENT RESEARCH

Almost all of my studies of cat sounds have revealed striking differences between the different sounds, but they have also revealed substantial variations in the melody within each sound category. The pitch (melody) may rise, fall, or first rise and then fall (or the other way around). The changes in melody may happen slowly or quickly. These are just a few examples of the changes that may occur.

When we phoneticians study human speech, we do not only investigate, as already mentioned, the vowels and consonants, but we also investigate the melody, rhythm and loudness—what we call the prosody (see pages 156–161). Although the cat has a very large vocal (acoustic) repertoire, cats cannot form as many different vowels and consonants as humans, and they do not have access to a large number of words, let alone sentences, so they do not have a grammar. It is possible that this is why cats vary their sounds so extensively, so that they can send different signals and express different wishes and needs.

The laboratory where I work—the Lund University Humanities Lab—is a very inspiring environment in which to work with many successful and nice colleagues. When I discussed my first studies of cat sounds with them, I got a lot of ideas, especially concerning the intonation in the communication between humans and cats. Moreover, many of my colleagues have cats themselves, and offered to take part in my studies. That encouraged me to formulate my questions and ideas into an application for a research project, and a private foundation (the Marcus and Amalia Wallenberg Foundation in Sweden) provided me with the financial support for a project that began in 2016 and will continue until 2021. A dream come true!

In my research project "Melody in Human–Cat Communication," I investigate the intonation in cat sounds in comparison to human speech together with my colleagues Dr. Robert Eklund and Dr. Joost van de Weijer. We came up with the name *Meowsic* by combing the word *meow* with *music*, so *meow-music*. Our main research questions are: What meaning does the melody have in cat sounds as compared to in human speech? Do cats use different melodies when they are content, sad, dissatisfied and angry? Do they learn to change the melody of their sounds in an attempt to be better understood by us humans?

The goal of our five-year research project is to find out how the melody in cat sounds varies when cats communicate with humans, and also how cats perceive human speech with its various vocal registers, melodies and speaking styles. The project consists of two parts.

In Study 1, we are recording sounds from thirty to fifty cats in different situations and investigating the variation in melody. We are primarily interested in recording sounds that cats produce when they interact with humans. Do cats' voices display different melodies when they are trying to express that they want something to eat, when they greet their humans, or when they are nervous or afraid? Does the melody vary with the degree of urgency, for instance when the cat is more or less hungry? If so, how do we humans perceive this difference in melody? According to our preliminary studies, the melody in cat sounds varies extensively and we humans can learn to perceive these variations so as to better understand the vocal signals of cats.

In Study 2, we investigate how cats perceive human speech. We will record multiple examples of human voices and speaking styles and play them back to cats to determine

1. Whether cats recognize the voices of their own humans;

2. Whether cats prefer a particular kind of melody or speaking style.

In order to better study the natural reactions of cats without putting them in stressful situations, we use low screens of our own construction, which we clean carefully to neutralize any odors, so that they can be brought into the cats' natural environments. Speakers are built into the screens, from which audio samples can be played. They also have video cameras to record the cats' reactions (for example, the movements of their ears, eyes, heads and bodies) when they hear the various examples of human voices and speaking styles.

The project is a pioneering work mainly in the field of linguistics (phonetics), but we are also consulting several Swedish and international experts from veterinary medicine, ethology (animal behavior) and general linguistics who specialize in human–animal communication. I also enjoy discussing our progress and our challenges with my colleagues at the Lund University Humanities Lab, and they offer their help gladly when we need expert assistance, for example in procedures for video recording. By being able to discuss every step in our studies with multidisciplinary experts—from preparation, through the research questions, and up to the results—we have a better chance of understanding the multifaceted sounds which cats use in their communication with us.

In addition to scientific discovery, our research has other positive side effects. A better understanding of the communication between cats and humans might lead to improved well-being for our feline companions, positively influence our interactions with pets, and also has the potential of improving the relationship between humans and animals in various areas—for example in animal breeding, animal therapy, veterinary medicine, and animal sanctuaries.

You will find more information about our project at our homepage, http://meowsic.info.

CAT-ASSISTED THERAPY

It has become commonplace that domestic animals support therapists in their work with patients. Horses, for example, have aided in therapy for depression or in work with mentally ill children. Hospitals employ specially trained dogs in treating chronically ill children and the elderly. Dogs and cats are increasingly used as companion animals for children who need a nonjudgmental friend while they practice their reading, for example, or for elderly people who would otherwise be alone.

Cats have some disadvantages, but also some advantages, when compared to dogs, who are much more frequently employed as therapy animals. Dogs are much easier to train and have better noses, capable of immediately perceiving tiny changes in scent. Cats, on the other hand, are often easier to keep. They do not need as much food, and one does not have to walk them several times each day. If we want to employ more animals as therapy and care assistants, it is very important that we humans can properly communicate with them and understand their signals, including their vocal signals.

Our current project, Meowsic, is especially concerned with the possibility of using cats as care or therapy assistants. Which characteristics does a cat have to possess in order to keep a human company, or even help with various problems? We can teach children which cat sounds are friendly and which are used as warnings. If a child knows that a low-pitched and dark (with acoustically low resonances) trill is a friendly greeting, they need not react with fear or aggression. In the end, it is very simple: if we want to understand our cats better—regardless of the situation—we just need to learn a little more about the signs they use to communicate and open ourselves up to them by listening very carefully. When we learn to better understand our cats and also to better communicate with them, we not only improve the well-being of the cat. We, too, are enriched.

APPENDIX

AUDIO AND VIDEO EXAMPLES OF SOUNDS PRODUCED IN DIFFERENT SITUATIONS

On my website, http://meowsic.info/catvoc, you will find an overview of the most common cat sounds presented in this book organized by title. These also depict the situations in which I recorded the relevant sounds. As I learn more about the various sounds of cats almost every day, the project continues, and the website is expanded regularly. That is why you may find more examples there than are provided here in this book. Above all else, I would like to show you the large phonetic range within each different sound category and hope that it will amuse you and that you will perhaps even recognize a few sounds from your own cat.

Sounds Produced with the Mouth Closed

1. *Purr(ing)*

Examples 1-4: Our cats Vincent, Donna, Rocky and Turbo made their first contributions to the study of cat sounds with their purring. All four videos show them purring calmly. At the beginning I was investigating the movement of their ribs, that is to say, how the rib cage went up as they inhaled and went down as they exhaled, which helps to inform us as to which phase of breathing corresponds to which parts of the purr.

The videos can be found on the website under "Purr(ing)" and the following titles: "Domestic cat Vincent purrs," "Domestic cat Donna purrs," "Domestic cat Rocky purrs," "Domestic cat Turbo purrs."

Example 5: The combination of purring, trilling and squeaking is so cute, especially when our Donna does it. An example can be found on the website under "Purr(ing)" and the title "The soft squeak embedded within the purr."

Example 6: Turbo sleeps and snores in his basket on my desk, and when I pet him, he starts to purr.

An example can be found on the website under "Purr(ing)" and the title "Snoring and purring."

Example 7: When Turbo wants my attention urgently, he often starts with a raw, hoarse meow or squeak. He meows until he gets it and then immediately starts to purr and tread. This video clip shows how he normally does this.

An example can be found on the website under "Purr(ing)" and the title "Demanding meows and purrs."

2. *Trill, Chirr, Chirrup, Grunt, Murmur*

Example 1: Donna trills softly as she asks nicely to be let outside.

An example can be found on the website under "Trill, Chirr, Chirrup, Grunt, Murmur" and the title "Donna trills by the door."

Example 2: Kompis trills by the kitchen window so as to get a treat. When I hear this friendly sound, I open the window and give him a little something to eat.

An example can be found on the website under

"Trill, Chirr, Chirrup, Grunt, Murmur" and the title "Kompis trills by the window."

Example 3: Vimsan is in heat and trills softly.

An example can be found on the website under "Trill, Chirr, Chirrup, Grunt, Murmur" and the title "Vimsan trills softly."

Example 4: Donna squeaks, chirrups and purrs happily on my lap when she wants to cuddle with me.

An example can be found on the website under "Trill, Chirr, Chirrup, Grunt, Murmur" and the title "Donna squeaks, trills and purrs on my lap."

Example 5: Turbo first meows and then grunts, which can be understood as a friendly request.

An example can be found on the website under "Trill, Chirr, Chirrup, Grunt, Murmur" and the title "Turbo meows and grunts."

Example 6: Turbo sleeps, but grunts or murmurs softly as I pet him.

An example can be found on the website under "Trill, Chirr, Chirrup, Grunt, Murmur" and the title "Turbo grunts in his sleep."

Sounds Produced with an Opening-Closing Mouth

1. *Meow*

   a. *Mew*

   Example 1: As already mentioned, Vimsan had a large wound on her rear leg when we found her. After being treated by the vet, she had to wear a cone collar for protection. Back then, she often mewed in this sad way. We still don't know if it was because she was in pain or because she wanted to get rid of her uncomfortable collar and be allowed to go outside again.

   An example can be found on the website in the sound category "Meow" under the title "Vimsan is mewing sadly."

   Example 2: Vimsan has come in from the garden to escape the rain and mews because she wants her food.

   An example can be found on the website under "Meow" and the title "Vimsan mews because she is wet and hungry."

   b. *Squeak*

   Example 1: Donna wants to play with me (or

wants me to open the door to the garden) and squeaks demandingly to summon me.

An example can be found on the website under "Meow" and the title "Donna squeaks demandingly."

Example 2: Donna squeaks, trills, and even produces a combination of trilling and squeaking when she wants to show me something urgently (often where the door to the garden, where she wants to be let out, is located), as though she wants to say "come on, follow me."

An example can be found on the website under "Meow" and the title "Donna squeaks and trills."

Example 3: Donna often combines soft and hoarse squeaking sounds when she wants to cuddle (after some squeaking she also mixes and combines several trills and purrs). An example can be found on the website under "Meow" and the title "Donna squeaks and purrs."

c. *Moan*

Example 1: Donna, Rocky and Turbo are sitting in their carriers in the waiting room of the vet moaning.

An example can be found on the website under

"Meow" and the title "Donna, Rocky and Turbo moan at the vet's."

Example 2: Kompis moans, meows, howls and growls in his carrier in the car on the way to the vet with the clear message: "Get me out of here!"

An example can be found on the website under "Meow" and the title "Kompis moans, meows, howls and growls."

d. *Meow*

Example 1: Rocky and Turbo meow because they have noticed that I am preparing shrimp in the kitchen and they would like to have some.

An example can be found on the website under "Meow" and the title "Rocky and Turbo meow to solicit food."

Example 2: Turbo meows in slow, hoarse tones when he urgently wants my care and attention (and when he gets it he often starts to trill and purr).

An example can be found on the website under "Meow" and the title "Turbo meows and purrs."

Example 3: Zoran, who lives with our friends Marie and Peter, meows until they open the basement door for him.

An example can be found on the website under "Meow" and the title "Zoran meows by the cellar door."

2. *Trill-Meow*

Example 1: Donna often trill-meows or trill-squeaks demandingly with a rising melody when she wants to play with me. With meows, trills and trill-meows Donna urges me to follow her because she wants to be let out into the garden.

An example can be found on the website under "Trill-Meow" and the title "Donna trill-meows and trill-squeaks."

3. *Howl, Yowl, Moan or Anger Wail*

Example 1: Kompis and an opponent howl a duet until the opponent leaves our garden in slow motion.

An example can be found on the website under "Howl, Yowl, Moan or Anger Wail" and the title "Kompis howls at intruder."

Example 2: Red (with the low-pitched voice) and an unknown opponent howl in an [oɪoɪoɪ]-like vowel pattern in our neighbor's garden.

An example can be found on the website under "Howl, Yowl, Moan or Anger Wail" and the title "Red and intruder howl."

Example 3: On a summer morning in Haapsalu, Estonia, I took a walk and saw this scene, in which two cats performed a duet of howling and screaming.

An example can be found on the website under "Howl, Yowl, Moan or Anger Wail" and under the title "Two cats howl and cry (scream) in Haapsalu."

Example 4: Vimsan walks up to Donna, and Donna answers with a brief hiss, which then becomes a longer howl. Vimsan growls softly before she withdraws.

An example can be found on the website under "Howl, Yowl, Moan or Anger Wail" and the title "Donna hisses and howls, Vimsan growls."

Example 5: Rivals Kompis and Teddy howl at each other, and then the smaller Teddy tries to creep away (in slow motion) through a balancing act on the garden bench.

An example can be found on the website under "Howl, Yowl, Moan or Anger Wail" and the title

"Kompis and Teddy are howling and moving in slow motion."

4. *Mating Call (Mating Cry)*

Tomcat Searching for a Female in Heat

Example 1: The tom Red wanders through our garden, marking fences and plants, rubbing himself against the garden furniture and meowing repeatedly. Maybe he is looking for a female in heat.

An example can be found on the website under "Mating Call (Mating Cry)" and the title "Red calls and meows."

Example 2: Because his wounds from a territorial dispute were being treated, our vet wanted to wait to neuter Kompis until everything was fully healed. I filmed him as he was producing trills and longing meows at our females (Donna and Vimsan) a couple times that spring.

An example can be found on the website under "Mating Call (Mating Cry)" and the title "Kompis calls for female cat."

Example 3: Here, one hears clearly that Kompis

meows at Donna in a different tone of voice—more low-pitched than he meows at me (more high-pitched and much brighter [with acoustically high resonances], almost like a kitten.)

An example can be found online under "Mating Call (Mating Cry)" and the title "Kompis meows at Donna and me."

Mating Calls and Other Sounds by Female Cats in Heat

Example 1: Vimsan is in heat and "treads" the floor with her hind legs with her rump in the air.

An example can be found on the website under "Mating Call (Mating Cry)" and the title "Vimsan coos and trills softly."

Example 2: If you listen carefully, you may be able to hear a few soft trills as Vimsan, who is in heat, rolls around on the floor.

An example can be found on the website under "Mating Call (Mating Cry)" and the title "Vimsan trills quietly."

Example 3: Vimsan, who is in heat, wanders restlessly around our house mewing softly and repeatedly.

An example can be found on the website under "Mating Call (Mating Cry)" and the title "Vimsan is in heat and mews softly."

I continue to find good examples of the songs of females in heat on YouTube. Some females combine trilling with loud moaning and meowing. Others call loudly, but somewhat differently. I have collected a few examples of other females in heat on my website.

An example can be found on the website under "Mating Call (Mating Cry)" and the titles "Female cat mating call 1" and "Female cat mating call 2."

Sounds produced with an open tense mouth

1. *Growl*

Example 1: Vimsan sits in her carrier at the vet's, sees a dog and starts to howl, though the howl becomes a deep growl.

An example can be found on the website under the category "Growl" and the title "Vimsan growls in her carrier."

Example 2: Vimsan growls softly to defend herself, as Donna spits and howls at her.

An example can be found on the website under "Growl" and the title "Vimsan growls softly."

2. *Hiss and Spit*

Example 1: Vimsan runs up to Donna and surprises her. She gets hissed at and then howled at for her trouble.

An example can be found on the website under "Hiss and Spit" and "Donna hisses and howls."

Example 2: Vimsan runs up to Donna, and Donna answers with a brief hiss, which then turns into a somewhat longer howl. Vimsan growls softly and finally withdraws.

An example can be found on the website under "Hiss and Spit" and the title "Hissing, howling and growling."

3. *Snarl, Cry or Pain Shriek*

Example 1: The spayed Vimsan defends herself against the still-unfixed Kompis. He follows her up an apple tree, but when he gets closer, the furious Vimsan snarls, hisses, spits and growls until he climbs back down.

An example can be found on the website under "Snarl, Cry or Pain Shriek" and the title "Vimsan snarls, hisses, spits and growls."

Example 2: Another encounter between Vimsan and Kompis in the apple tree.

An example can be found on the website under "Snarl, Cry or Pain Shriek" and the title "Vimsan growls, snarls and spits."

4. *Chirp and Chatter (Prey-Directed Sounds)*

   i. *Chatter (Teeth Chattering)*

   Example 1: Our Rocky sits at the kitchen table, chatters and chirps at a bird at the window, jumps down and runs to the window.

   An example can be found on the website under "Chirp and Chatter (Prey-Directed Sounds)" and the title "Rocky chatters and chirps."

   ii. *Chirp*

   Example 1: Turbo sits by the kitchen window and chirps at a bird. After that, he loses interest and jumps, trilling, off the windowsill.

An example can be found on the website under "Chirp and Chatter (Prey-Directed Sounds)" and the title "Turbo chirps by the window."

Example 2: Donna often sits on the windowsill in the kitchen and chirps at birds (sometimes she trills intermittently).

An example can be found on the website "Chirp and Chatter (Prey-Directed Sounds)" and the title "Donna chirps by the window."

iii. *Tweet*

Example 1: Rocky has a large vocabulary of chirps. He combines chirp sounds with soft tweeting.

An example can be found on the website under "Chirp and Chatter (Prey-Directed Sounds)" and the title "Rocky tweets at birds."

iv. *Tweedle*

Example 1: Rocky can also combine chirping and twittering with long tweedling, in many melodic variations and with tremolo.

An example can be found on the website under "Chirp and Chatter (Prey-Directed Sounds)" and the title "Rocky tweedles."

TABLE 2: OVERVIEW OF THE MOST COMMON CAT SOUNDS

This table summarizes all of the sound types and their corresponding phonetic characteristics. Here you will find the phonetic characteristics for the sounds and their subcategories, such as the type of articulation (the position or movement of the mouth), the register of the voice (voiced or voiceless, high or low pitch [melody]) and the typical phonetic transcription, as well as some additional comments.

Vocalization Type	*Subcategory*	*Articulation (Mouth)*	*Voice*
Meow	Mew	Opening (open)	Voiced, very high-pitched/bright
Meow	Squeak	Opening	Voiced, high-pitched, bright, hoarse, raspy
Meow	Moan	Opening-closing	Voiced, often falling tone
Meow	Meow	Opening-closing	Voiced, often rising-falling (but much variation)
Trill-meow	Trill-meow	Closed-opening (-closing)	Voiced, rising tone
Trill	Chirrup, coo (weaker), chirr (brighter/shriller)	Closed, airflow through the nose	Voiced, high-pitched, bright, rising tone
Trill	Grunt, murmur	Closed, airflow through the nose	Voiced, lower/darker, often level or falling tone
Growl	Growl, snarl	Slightly open	Voiced, very low
Hiss	Hiss	Open	Voiceless
Hiss	Spit	Open	Voiceless

Phonetic Category	*Typical Phonetic Transcription*	*Comments*
High-pitched meow, often with [i], [ɪ], [e] and [u] vowel(s)	Often [mi], [wi] or [mɪu]	Isolation, or soliciting sound or call (often by kittens)
Hoarse, raspy, nasal, bright often short mew-like sound with [ɛ] or [æ]	Often [wæ], [mɛ] or [ɛʊ]	Isolation, or soliciting sound or call (often by adult cats)
Darker meow, often with [o] or [u] vowel(s)	Often [mou] or [wuæu]	Discontent or distressed sound
Combination of several vowels resulting in the typical [iau] sequence	Often [miau], [ɛau] or [wɑːʊ]	The most common human-directed attention-soliciting sound
A trill directly followed by a meow	Often [bʀ̃iuw], [br̃ːmiau], [mhr̃iauw] [mhr̃ŋ-au] or [whr̃ːau]	Common human-directed sound to solicit attention
Trilling sound sounding a bit like a Scottish rolled [r], but nasal and probably produced farther back in the mouth	Often [mʀ̃ːh] or [mːr̃ːut]	*The* friendly greeting or calling sound
Trilling sound sounding a bit like a Scottish rolled [r] or a guttural (French) rolled [ʀ], but nasal, often hoarse and probably produced farther back in the mouth	Often [m̰ː] or [bʀ̃ː]	Friendly sound often used in greeting and confirmation contexts, murmur is sometimes produced as a pure nasal sound [m] (without trilling)
Very low extended trill, sometimes beginning with a creaky [m̰]	Often [gʀː], [ʀː] or creaky trilling [ɹ̰ː], [ʌ̰ː] or [m̰ʀː]	Warning sound
Dark (back) or bright (front) fricative (with audible friction)	Often [fːhː], [çː], [ʃː], [ɧ̰] or [ʂ̰ː]	Warning sound
Often dark (back) or bright (front) africate (stop + fricative)	Often [t͡ʂ̰ː], [kh̯ː] or [k͡ʃː]	Explosive warning sound

Vocalization Type	*Subcategory*	*Articulation (Mouth)*	*Voice*
Howl	Howl, mating call	Open (slightly opening-closing)	Voiced, melody rises and falls in repeated patterns
Growl-howl	Growl-howl	Closed-opening-closing	Voiced, melody rises and falls between very low-pitched (growl) and very bright (howl)
Snarl (cry, scream)	Snarl (cry, scream)	Tense and open	Voiced, often hoarse and harsh, level or falling melody
Mating call (mating cry)	Mating call (mating cry)	(Closed-) opening-closing	Voiced, often final rise in melody
Purr	Purr	Closed, airflow mostly through the nose	Probably mainly voiceless, but regularly vibrating, extremely low-pitched (20 Hz)
Chirp and chatter	Chatter, teeth chattering	Open	Voiceless
Chirp and chatter	Chirp	Open	Voiced, often loud short sequences
Chirp and chatter	Tweet	Open (slightly closing)	Voiced, often soft short sequences
Chirp and chatter	Tweedle	Open (slightly closing-opening)	Voiced, soft longer sounds

Phonetic Category	*Typical Phonetic Transcription*	*Comments*
Combination of two or more extended vocalic sounds, including [ɪ], [ɨ], [j], [ɤ], [au], [ɛɔ], [aw], [oɪ] and [ɑo]	E.g. [awoɪːɛɔː], [jiɨɛɑw] or [ɪːauauauauauawawaw]	In the same frequency region (band?) as human baby crying
Combination of growl and howl with substantial rises and falls in the melody	E.g. [gʀːawɪjɑoʀː]	Warning sound
Short, often loud vocalic sounds	Often [a], [æ], [aʊ] or [ɛo]	The sound of anger, pain or warning, used in fighting or when in pain
Long stressed vowels, often begins with [w] or trilling consonant sound	Often sequences of [wa͡ːuw], [r̃ːɪːa͡uː], [mhr̃ːwaːoːuːɪː] and [ʀ̃ːwːuːaːu]	Often "concerts" with two rivals or a single female in heat lasting for hours, especially on spring nights
Soft, extended, very low-pitched breathy vibrating sound, e.g. [ʀ̃] or [r̃], often combined with soft [h] consonants, produced during both ingressive and egressive airflow	E.g. [↓hːr̃-↑r̃ ːh-↓hːr̃-↑r̃ːh]	More likely to signal "I do not pose any threat" than "I am content," but we still don't know exactly how cats purr
Iterative consonants, often sounding like [k] or [ʔ] (glottal stop)	Often [ʔ ʔ ʔ ʔ] or [k̥⁼ k̥⁼ k̥⁼ k̥⁼ k̥⁼ k̥⁼]	Often prey directed
Often initial glottal stop [ʔ] followed by short vowel, often [ɛ], [e] or [ə]	Often [ʔə], [k̥⁼e] or [ʔɛʔɛʔɛ]	Often prey directed
Soft chirp without initial [ʔ], sometimes with initial [w], vowels often [i], [ɪ], [ɛ] or [u]	Often [wi] or [ɦɛu]	Often prey directed
Long extended chirp or tweet, often with voice modulation like tremor or quaver	E.g. [waɛəɥə] or [ʔəɛəɥə]	Often prey directed

TABLES WITH PHONETIC SYMBOLS

TABLE 3: VOWELS

This table presents the vowels that I have observed in cat sounds. In addition, there are some vowels that I have not yet heard, but which I assume cats to be capable of producing. Most vowels can be either short or long. When long vowels occur in a phonetic transcription in the book, they are followed by a length symbol [ː]. English also distinguishes between long and short vowels, for example the short *oo* in *good* [ʊ] and the long *oo* in *school* [uː].

Phonetic Symbol	Example words (intended sound in **boldface**)	Examples in cat sounds	Description
[a]	German: K**a**tze, K**a**ter; first part of the diphthong [aɪ] in m**i**ce	[miaʊ]	open front unrounded vowel
[ɑ]	RP English: h**a**rd	[wɑːʊ]	open back unrounded vowel
[ɐ]	Australian English: r**u**n	(not yet observed)	near open central unrounded vowel
[ɛ]	b**e**d	[ɛaw]	open-mid front unrounded vowel
[æ]	c**a**t, h**a**nd	[wa æh æh]	near-open front unrounded vowel
[ə]	**a**bout	[ʔɛ ʔə]	open-mid front unrounded vowel (Schwa)
[e]	German: g**e**guckt, f**e**hlen; first part of the English dipthong [eɪ] in d**ay**	[meːʊ]	close-mid front unrounded vowel
[i]	m**ea**t	[miu]	close front unrounded vowel
[ɪ]	f**i**sh	[mɪ-ɑːou]	near-close, near-front unrounded vowel
[ɨ]	Polish: s**y**n (son)	[jiɨɛɑw]	close central unrounded vowel
[ɔ]	h**o**rse	[ɛɔ]	open-mid back rounded vowel
[ʌ]	r**u**n	[ʌ̰ː] (e.g. growling)	open-mid back unrounded vowel
[o]	Scottish English: n**o**; first part of the American English diphthong [oʊ] in s**o**	[oːɪoːɪoːɪoːɪ]	close-mid back rounded vowel
[ɤ]	Estonian: s**õ**na (word) (unrounded [o])	(not yet recorded)	close-mid back unrounded vowel
[œ]	German: H**ö**lle (rounded [ɛ])	(not yet observed)	open-mid front rounded vowel
[ø]	German: L**ö**we (rounded [e])	(not yet observed)	close-mid front rounded vowel
[ʊ]	g**oo**d	[miaʊ]	near-close central rounded vowel
[u]	sch**oo**l	[miu]	close back rounded vowel
[ʏ]	German: Sch**ü**ssel (rounded [ɪ])	(not yet observed)	near-close near-front rounded vowel
[y]	German: s**ü**ß (rounded [i])	(not yet observed)	close front rounded vowel

TABLE 4: CONSONANTS

In this table, I have compiled the consonants that I have been able to hear in cat sounds. In addition, there are a few consonants that I have not heard, but which cats should be able to produce.

Phonetic Symbol	Example words (intended sound in boldface)	Examples in cat sounds	Description
[ʔ]	u**h**-o**h**	[ʔɛʔɛʔɛ]	glottal stop
[b]	**b**ut	[br̃ːiau]	voiced bilabial stop (or plosive)
[ç]	German: Mil**ch** (milk)	[çː] (e.g. hissing)	voiceless palatal fricative
[ɕ]	Swedish: **k**ärlek (love)	[ɕː] (e.g. hissing)	voiceless alveolopalatal fricative
[f]	**f**un	[fːhː]	voiceless labiodental fricative
[g]	**g**o	[gʀː]	voiced velar stop (or plosive)
[h]	**h**ouse	[fːhː]	voiceless glottal fricative
[ɦ]	a**h**a!	[ɦɛu]	voiced glottal fricative
[j]	**y**es	[jɪɨɛɑʊw]	voiced palatal approximant
[k]	**c**at	[k̩⁼ k̩⁼ k̩⁼ k̩⁼ k̩⁼]	voiceless velar stop (or plosive)
[l]	**l**eft	(not yet recorded)	voiced alveolar lateral
[m]	**m**ouse	[mhr̃ː]	voiced bilabial nasal
[n]	**n**ose	(not yet recorded)	voiced alveolar nasal
[ŋ]	si**ng**	[mhr̃ŋ-au]	voiced velar nasal
[p]	**p**en	(not yet observed)	voiceless bilabial stop (or plosive)
[pf]	German: To**pf**	(not yet observed)	voiceless labiodental affricate
[r]	**r**at (Scottish rolled *r* with vibrating tongue)	[mːr̃ːut]	voiced alveolar trill
[ɹ]	**r**at	[ɹ̰]	voiced alveolar approximant
[ʀ]	**r**at (French rolled *r* with vibrating uvula)	[gʀːawɪjɑoʀː]	voiced uvular trill
[ʂ]	Swedish tö**rs**t (thirst)	[t͡ʂ̰]	voiceless retroflex fricative
[ʃ]	**sh**e	[kʃːt]	voiceless postalveolar fricative
[t]	**t**in	(observed only in affricates)	voiceless alveolar stop (or plosive)
[t͡ʃ]	**ch**ur**ch**	[t͡ʃ]	voiceless alveolar affricate
[w]	**w**e	[whr̃ːau]	voiced bilabial approximant
[ɥ]	French: h**u**it (eight)	[wəɛə̣ɥə]	voiceless velar approximant

TABLE 5: OTHER PHONETIC SYMBOLS

In addition to vowels and consonants, I have used some special symbols to describe some phonetic characteristics or clues that I have found in cat sounds.

Phonetic Symbol	Example words (intended sound in **boldface**)	Examples in cat sounds	Description
[ː]	sch**oo**l	[wɑːʊ]	Length symbol (preceding symbol is pronounced long)
[k̟]	i**k**i	[k̟⁼ k̟⁼ k̟⁼ k̟⁼ k̟⁼]	Pronounced farther front
[k⁼]	i**ck**i	[k̟⁼ k̟⁼ k̟⁼ k̟⁼ k̟⁼]	Unaspirated (without any burst of breath following)
[a͡u]	m**ou**se	[a͡ʊ]	Pronounced together (like a diphthong)
[~]	French: b**õ**n (good)	[r̃] (e.g. trilling)	Nasal or nasalized
[m̰]	(an [m] with creaky voice quality)	[m̰ː] (e.g. trilling)	Creaky (pronounced with creaky voice quality or vocal fry)
[↓]	North Swedish: ja (yes) (pronounced during inhalation)	[↓hːr̃-↑r̃ːh-↓hːr̃-↑r̃ːh]	Ingressive (pronounced during inhalation)
[↑]	yes (pronounced during exhalation)	[↓hːr̃-↑r̃ːh-↓hːr̃-↑r̃ːh]	Egressive (pronounced during exhalation)

GLOSSARY— IMPORTANT TECHNICAL TERMS

Acoustic: of or relating to sound.

Affricate: *plosive* sound followed by a *fricative*, such as in *jacket* and *cheese.*

Alveolar: a sound made with the tongue on the top of the mouth, specifically on the ridge behind the teeth, such as *s, t, d* and *z* in *stop, debt, zoo.*

Approximant: a *consonant* which is approaching (approximating) the articulation of a vowel, i.e. it is produced with a wider (larger) distance between the articulators than in *fricatives* (which have a more narrow passage), but a smaller distance than in vowels. Examples include *l*, *r* and *w* in *less*, *rest* and *west*.

Aspirated: a stop sound with a clearly audible exhalation, such as *p* in *pin*.

Auditory: something to do with one's sense of hearing; sense something acoustically.

Bilabial: a *consonant* formed with both lips, such as *b*, *m* and *p* in *bed*, *man* and *spit*.

Consonant: sound produced by a narrowing of the vocal tract; one distinguishes between *voiced* (e.g. *b*, *n*, *l*) and *unvoiced* (e.g. *f*, *k*, *s*) consonants, as well as by the site of production (lips, gums, tongue, etc.), as well as by the kind (stop, *fricative*, *approximant*, *nasal*).

Diphthong: a double-vowel, i.e. a vowel within a single syllable which changes from one distinct quality to another.

Egressive: a sound produced with air streaming out through the mouth; the opposite of *ingressive*.

Electromagnetic Articulography: procedure by which the speech movements can be (clinically) investigated. It can be used to analyze the movement of the tongue during speech, for example.

Fricative: either a *voiced* or a voiceless *consonant* produced with frication noise due to a narrow passage in the vocal tract, such as the first letters in *sip*, *ship* and *zip*.

Glottal: glottal sounds are produced in the glottis, which is the name for the opening between the vocal folds (vocal cords).

Ingressive: the opposite of *egressive*, the sound is produced when air is inhaled, that is to say against the normal flow of speech. An example is when one says "huh!" to express surprise.

Kneading: a rhythmic stepping used by kittens to stimulate their mother's production of milk. Domestic and wild cats both also knead later in life, for example when they want to set up a bed.

Labiodental: *consonants* articulated with the lips and teeth, such as the *f* in *fun* and the *v* in *van*.

Lateral: a sound produced by raising the tip of the tongue against the palate (the roof of the mouth) so that the airstream flows past one or both sides of the tongue, such as the *l* in *lots* and *lean*.

Nasal: a sound or a *consonant* produced when the air is expelled in whole or part through the nose, such as the *m* in *many*, the *n* in *never* and the *ng* in *sing*.

Palatal: a sound produced using the tongue and the hard palate in the middle of the roof of the mouth, such as the *y* in *yes*.

Pheromone: a scent-carrying substance that carries information between members of a species and which triggers a reaction (e.g. calmness or excitement) in its recipient.

Phoneme: the smallest semantically significant unit of speech, *vowel* and *consonant* sounds. By changing a phoneme one changes the meaning of a word, such as when one changes the *b* in *beg* to the *p* in *peg*, the *h* in *house* to the *m* in *mouse*, or the *o* in *mode* to the *a* in *made*.

Phonetician: a researcher who concerns themselves with *phonetics*.

Phonetics: the branch of human inquiry that concerns itself with the characteristics and meaning of the sounds of human speech, for example how and where a sound is produced, how is it received by the listener and processed. There are therefore physical, physiological and psychological aspects.

Plosive: also called a stop *consonant*; a *voiced* or *unvoiced consonant* is formed when the stream of breath has been stopped and then starts, such as the *p* in *partner* or the *t* in *trap*.

Postalveolar: a place of articulation. The sound, or *consonant*, is produced somewhat farther back in the mouth than with an *alveolar*, such as *ship*, *vision* or *chip*.

Prosody: all of the characteristics of speech that are not connected to a particular sound *(phoneme)*, but to everything that is pronounced, including the emphasis of a word or syllable, the tempo, breaks in speech, accent, intonation, etc.

Retroflex: a sound produced using a bent-back tongue, such as *d* in Indian English *bird*.

Semivowel: sounds which are neither clearly a *vowel*

nor clearly a *consonant*. Examples include the *y* in *yes* and the *w* in *was*.

Sibilant: usually a *fricative*, produced with the alveolar or the soft front of the gums, and can be either *voiced*, like the *z* in *zip*, or voiceless, like the *s* in *sip*.

Tactile: investigate or experience using the sense of touch.

Transcription: the representation of speech in a written form. In phonetic transcription, sounds are conveyed using a phonetic alphabet, which indicates how they should be pronounced.

Tremolo: shaking, quavering voice; a term used frequently in regard to singing.

Trill: a class of *consonants* where the tongue tip or the uvula is brought into trembling, vibrating motion, for example a rolled *r*, like when one expresses that it is cold by saying *brrr*, or in some Scottish dialects.

Unvoiced: a sound produced without any vibrations of the vocal folds, which are so far apart that air flows through them without hindrance and is blocked only farther forward in the oral cavity. Some, but not all, *fricatives* and *plosives* are unvoiced, e.g. *f*, *t* and *h*.

Uvular: a sound produced with the back of the tongue at or near the uvula. There are no uvular sounds in English.

Velar: a sound produced with the back part of the tongue against the soft palate (velum), such as *ng*, *c* and *g* in *ring*, *scope* or *get*.

Voiced: a sound that is produced using the vocal folds, which vibrate. All *vowels* are voiced (unless they are whispered), as are some *consonants*, such as *l*, *d* or *v*.

Vowel: a *voiced* sound in which the air is expelled through the mouth without hindrance. Examples of English vowels are *a*, *e*, *i*, *o* and *u*.

WORKS CITED AND FURTHER READING

Bradshaw, J. W. S. (2013). *Cat sense: The feline enigma revealed*. London, England: Penguin Books.

Brown, K. A., Buchwald, J. S., Johnson, J. R., & Mikolich, D. J. (1978). Vocalization in the cat and kitten. *Developmental Psychobiology, 11*(6), 559–570.

Clark, M. R. (2016). *Pussy and her language*. Wentworth Press. (Original work published 1895)

Darwin, C., & Ekman, P. (1998). *The expression of the emotions in man and animals* (3rd ed.). London, England: HarperCollins.

Dexel, B. (2014). *Birga Dexel's clickertraining für katzen.* Stuttgart, Germany: Kosmos-Verlag.

Leyhausen, P. (2005). *Katzenseele: Wesen und sozialverhalten.* Stuttgart, Germany: Franckh-Kosmos.

McComb, K., Taylor, A. M., Wilson, C., & Charlton, B. D. (2009). The cry embedded within the purr. *Current Biology, 19*(13).

McNamee, T. (2017). *The inner life of cats.* New York, NY: Hachette.

Moelk, M. (1944). Vocalizing in the house-cat; a phonetic and functional study. *The American Journal of Psychology, 57*(2), 184.

Ohala, J. J. (1994). The frequency code underlies the sound symbolic use of voice pitch. In L. Hinton, J. Nichols, & J. J. Ohala (Eds.), *Sound symbolism* (pp. 325–347). Cambridge: Cambridge University.

ACKNOWLEDGMENTS

I find it almost unbelievable: I, a Swede, who had only written very little in German before this book, was able to write a book about my major research interest and my favorite hobby, cat sounds—in German! I would never have been able to do so without the help of many people. If Bettina Stimeder from my Austrian publisher Ecowin Press had not called me up and asked me if I was interested in writing a book about cat sounds, I would still be dreaming of such a book. Without my German editors, Friederike Thompson and Silke Martin, nobody would be able to read the German version of the book, as I write a terribly bad German!

I was thrilled when I heard that the book was being translated into English and when I was asked to help with the translation. This was another challenge for me. It is not an easy task to try to find the corresponding idiomatic expressions or the correct phonetic terminology, let alone clear examples, in a foreign language (for instance the vowel system of English is so different from the German or Swedish ones). I owe my colleague Eva Liina Asu-Garcia at University of Tartu, Estonia, a huge thank-you for helping me with the phonetic parts of the English translation.

My PhD supervisor, fellow phonetician and friend Dr. Per Lindblad, read the entire German manuscript carefully and discussed it with me for hours; for that, I am eternally grateful. It was a lot of fun to discuss my text with him and he made a lot of suggestions for improvements. The phonetic parts of the book were often inspired by his own books and compendiums. Unfortunately, they are available only in Swedish; otherwise I would recommend them to anyone who would like to learn a little more about phonetics. Dr. Gilbert Ambrazaitis, a German phonetician, also read most of the sections on phonetics and helped me to find the right technical terms in German. Thanks, Gilbert! I would also like to thank all of my colleagues at the department of linguistics and the division of logopedics, phoniatrics and audiology at Lund University, as well as all the people who came to listen to my lectures and talks on cat sounds and then discussed them with me afterward. I would especially like to thank my collaborators on multiple studies of cat sounds, and for my project "Melody in Human–Cat Communication," Robert Eklund and Joost van de Weijer, as well as all of my colleagues at the Lund University Humanities Lab. I owe the Marcus and Amalia Wallenberg Foundation an enormous thank-you for choosing to support my project, as well as the many nice reporters and journalists who interviewed me about my work—often in German or English—and helped me explain it to a lay audience. A warm thank-you also goes to Birga Dexel and Tanja Warter, whom I met in 2017 in Bregenz, Austria, at the Animalicum Conference. I learned so much about cat medicine and cat behavior from them, and they gave me the courage to communicate my little contribution to cat research in German.

Thanks, too, to all the people who sent me mail with video or audio recordings of their cats. These additional recordings of cats really made it possible for me to write about the sounds of cats in general and not only the sounds of my own cats. So thanks, too, to all the cats whose sounds I got to listen to and study.

Finally, thanks to the main characters in my book, my wonderful cats Donna, Rocky, Turbo, Vimsan and Kompis, as well as our neighbor cat Graywhite. They taught me almost everything I know about cat sounds.

And of course, to my husband, Lars, with whom I share the joy and happiness that our cats bring us every day.

Many, many thanks to all of you—I know how lucky I am.